Deserts

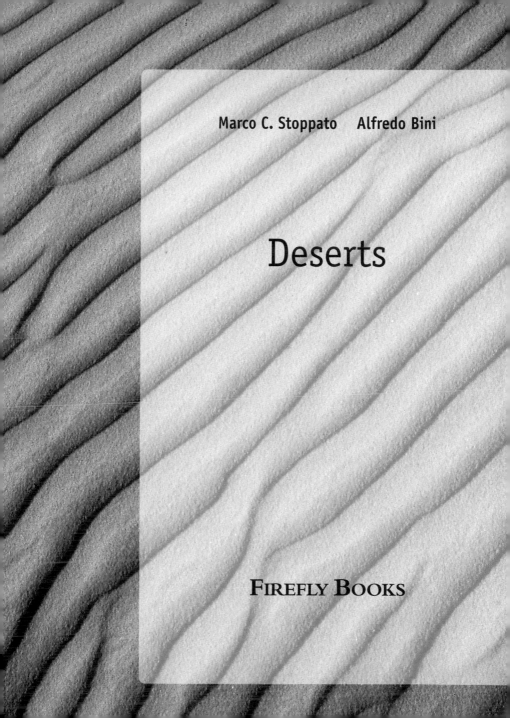

Marco C. Stoppato   Alfredo Bini

Deserts

FIREFLY BOOKS

A FIREFLY BOOK

Published by Firefly Books Ltd. 2003

First printing 2003

**Publisher Cataloging-in-
Publication Data (U.S.)**

Stoppato, Marco
     Deserts : A Firefly guide / Marco
C. Stoppato – 1st ed.
[256] p. : col.ill., photos. ; cm.
Includes bibliographical references
and index.
Originally published as *Deserti*,
Italy: Mondadori, 2001.
Summary: A comprehensive guide to
deserts on five continents including
formation, location, structure, dunes
and soil.
ISBN 1-55297-669-6 (pbk.)
1. Deserts. I. Bini, Alfredo. II. Title
551.415   21
GB611.S76   2003

**National Library of Canada
Cataloguing in Publication Data**
     Deserts: a Firefly guide / Marco
C. Stoppato, Alfredo Bini; translated
by Linda M. Eklund.

Translation of *Deserti*.
Includes bibliographical references
and index.
ISBN 1-55297-669-6
1. Deserts I. Bini, Alfredo
II. Eklund, Linda M. III. Title

GB611.S76   2003
551.41'5   C2002-903838-3

Translated by Linda M. Eklund
Science content advisor: Dr. Steven
H. Williams

Published in Canada in 2003 by
Firefly Books Ltd.
3680 Victoria Park Avenue
Toronto, Ontario M2H 3K1

Published in the U.S.A. in 2003 by
Firefly Books (U.S.) Inc.
P.O. Box 1338
Ellicott Station
Buffalo, New York 14205

Printed in Spain by
Artes Gráficas Toledo S.A.U.
D.L. TO: 581-2003

CONTENTS

Symbols

Annual Precipitation

less than 25 mm (1 in)

between 25–50 mm (1–2 in)

between 50–100 mm (2–4 in)

between 100–150 mm (4–6 in)

between 150–250 mm (6–10 in)

Surfaces

mostly sandy desert surface

mostly rocky desert surface

highly eroded, arid terrain with no soil (desert zone in transition)

CLIMATE

Overleaf: A satellite image of a cyclone whose eye is clearly visible

Below: A mass of clouds over the Atlantic Ocean near the Azores archipelago.

Bottom: A dolomite landscape and below it, the Okawango delta in Botswana.

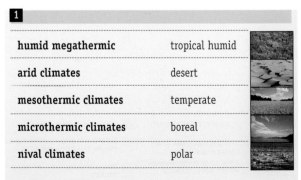

Climate is one of the primary factors that contribute to determining the morphology of an environment or region. Many factors converge to establish a climate and are used to classify it, including annual irradiation, the distribution of water masses and mountains, latitude and altitude. Temperature, precipitation and wind flows are also elements that determine climate. In fact, the development of vegetation and animal life is as bound to climate as is the shape assumed by a landscape. Thus, climate regulates humankind's ability to establish itself in specific places around the globe and can be thought of as the body of atmospheric conditions that manifest themselves and influence a particular region in a given geologic era. The distribution of native vegetation also plays a part in defining climatic regimes, but it represents a climatic effect rather than a cause, and is regulated by the soil types from which plants take their nutrition. Given the enormous variability of the elements involved in categorizing climates, several classifications proposed since 1884 take different factors into consideration.

Climatologist Wladimir Koppen published a classification of climates in 1936 based primarily on rainfall and temperature but also on the nature of various vegetal formations. Koppen then defined five major climate groups, summarized in Table 1.

The five groups include two or more **climate types** each, which in turn present additional corresponding **biomes** characteristic of various land masses (rainforest, savanna, desert, Mediterranean scrub, conifer forests, tundra and so forth). Though several other classi-

1

humid megathermic	tropical humid
arid climates	desert
mesothermic climates	temperate
microthermic climates	boreal
nival climates	polar

fications based on different parameters exist, they are generally derived from Koppen's perceptions. The great mass of data on daily temperature readings collected worldwide for decades favors a classi-fication based on the temperature of the atmosphere at ground level. An analysis of monthly averages identified three principal climate groups, summarized in Table 2.

The large quantities of available data make possible a classification based on precipitation in relation to geographic position.

Above: A colorful example of Mediterranean scrub in Corsica.

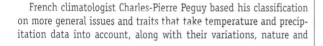

2

low-latitude climates (average monthly temperature never below 18°C (64°F))	**no winter**
middle-latitude climates	**winter and summer seasons**
high-latitude climates (average monthly temperature never above 10°C (50°F))	**no summer**

French climatologist Charles-Pierre Peguy based his classification on more general issues and traits that take temperature and precipitation data into account, along with their variations, nature and

Below: This map illustrates the global distribution of natural vegetation.

11

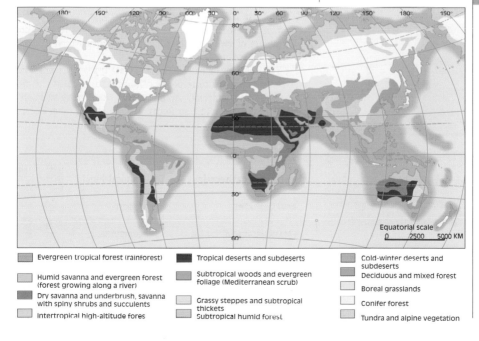

Evergreen tropical forest (rainforest)

Humid savanna and evergreen forest (forest growing along a river)

Dry savanna and underbrush, savanna with spiny shrubs and succulents

Intertropical high-altitude fores

Tropical deserts and subdeserts

Subtropical woods and evergreen foliage (Mediterranean scrub)

Grassy steppes and subtropical thickets

Subtropical humid forest

Cold-winter deserts and subdeserts

Deciduous and mixed forest

Boreal grasslands

Conifer forest

Tundra and alpine vegetation

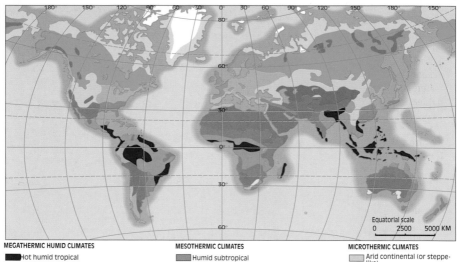

MEGATHERMIC HUMID CLIMATES
- Hot humid tropical
- Hot tropical with a dry season

ARID CLIMATES
- Hot semiarid tropical
- Tropical and subtropical desert
- Cold desert

MESOTHERMIC CLIMATES
- Humid subtropical
- Subtropical with dry winter
- Subtropical with dry summer (Mediterranean)
- Cool oceanic temperate (humid)
- Cool continental temperate (subhumid)
- Cool temperate with dry summer

MICROTHERMIC CLIMATES
- Arid continental (or steppe-like)
- Cold-climate forests (or Taiga)

NIVAL CLIMATES
- High-altitude cold (above the tree line)
- Subpolar
- Polar

Above: A modern pluviograph (rain gauge).

Top: This map illustrates the global distribution of different climate types.

DISTRIBUTION BY LATITUDE OF CLIMATIC ZONES			
equatorial humid zone	10°	N	10° S
windward tropical coasts	5°–30°	N	e S
tropical deserts	10°–35°	N	e S
middle-latitude deserts and steppes	30°–50°	N	e S
subtropical humid regions	25°–45°	N	e S
middle latitude western coasts	35°–65°	N	e S
arctic and polar deserts	60°–90°	N	e S

3

inland climates	of the great Antarctic and Greenland ice caps
cold climates	high latitudes
temperate climates	middle latitudes
Mediterranean and tropical climates	low latitudes

the manner in which they manifest. This classification then accords with the astronomical zones circumscribed by the polar circles and the tropics, and Peguy thus defines four zonal climates (shown in Table 3). A fifth category known as "azonic" does not depend upon latitude and can be added to this classification; it includes the *arid climate*.

More general classifications take additional geographic parameters and features into consideration defining as follows:

- equatorial climates
- tropical climates
- subtropical climates
- temperate climates
- hot-desert climates
- cold-desert climates
- arctic climates

Alternatively, five basic climate types can be ascertained from statistical data and a mathematical analysis of the average duration of given temperatures and rainfall:

- humid tropical climates
- dry climates
- hot humid temperate climates
- cool humid temperate climates
- polar climates

All these proposed classifications indicate geographic areas with distinct meteorological regimes. The morphological characteristics of each territory—those that define its landscape—are tied strictly to these regimes.

The very nature of the Earth's soil depends upon climate. Soil plays a fundamentally important role in the evolution of an environment (and will be discussed further in the chapter that starts on page 66). Soil formation begins with the disaggregation and minute splintering of rocks that is generally caused by atmospheric agents (and thus by climatic conditions) and by a whole series of chemical and biological changes. This early product of weathering is actually the regolith, which has no organic component.

The decisive climatic factors in **pedogenesis**, or soil formation, are **water** in the form of precipitation, **temperature** and **wind.** Indeed, biological and chemical activity would be impossible without water. Mineral salts in the terrain are eroded, dissolved and made inter-reactive by water,

Below: A sand dune in the Libyan Desert.

Bottom: The effects of exfoliation on rocks in Egypt's Western Desert, otherwise known as the White Desert.

giving origin to compounds that can be assimilated by plants and animals.

Temperature influences chemical activity in that ice blocks it, cold slows it down, and heat favors it. Mechanical weathering, however, is greatly facilitated by ice-wedging under repeated freeze-thaw cycles.

Beyond promoting the evaporation of moisture, wind transports

Clockwise: A view of the Amazonian rainforest in Venezuela; ice crystals; mosses in Iceland; an Arctic reindeer in Norway's Svalbard Islands; and a typical Arctic panorama.

surface material, causing erosion in some situations and accumulation in others. Soil, in the final analysis, is where the continual re-elaboration of mineralogical and organic elements occurs (up to 10 billion microbes can be found in every gram of soil) and where the processes of physical and biochemical alteration that are indispensable to plant and animal life are repeated.

Every climate type can thus be associated with a well-defined geographic region whose characteristics derive from the meteorological elements linked to it. A cold climate, for example, is banded by the 4°C (39°F) isotherm in the western hemisphere (south of that line the climate is temperate) and involves distinct geographic regions—arctic, subarctic and cold oceanic.

The temperate climate features balanced zones with summers and winters, moderate temperatures and precipitation around 1,500 mm (about 60 in) a year. Examples include the climate of Western Europe and North America's middle latitudes. Depending upon the number of hot and humid months that recur during the year, the humid tropical climate can be subdivided into attenuated tropical, medium tropical and hypertropical. The latter is responsible for the existence of rainforests like those in the Amazon basin.

And finally, whether hot or cold, we find arid climates defined in every classification and emphasized wherever deserts have formed. The concept of **aridity**, so intimately connected to deserts, is actually independent of temperature. Some arid zones are characterized by high temperatures. The Sahara Desert, for example, registers average annual temperatures above 18°C (64°F) and average temperatures no lower than 26°C (79°F) in the hottest month. On the other hand, there are cold deserts, such as the Gobi in winter and the polar deserts of Antarctica. The basic parameter that defines aridity is the quantity of precipitation. The maximum annual rainfall is 250 mm (10 in) a year for an area to be defined as a desert, but this number may drop much lower in some areas. In some extreme localities, such as Chile's Atacama Desert, decades can pass with no rain at all. Desiccated regions far from ocean water masses, and thus more landlocked, are extremely poor in humidity and intensely arid as a result. Precipitation does not exceed 120 mm (4.7 in) yearly, and rains are often concentrated into a single, violent and short downpour. In the Sahara's most extreme areas, an exceptional

rain event where the annual rainfall is discharged in a single storm is repeated with a frequency of three to four times every 30 years (see Table 4). Yearly precipitation levels between 250–400 mm (10–16 in) define a semiarid or pre-desert zone.

The lack of water is responsible for the lack of soil and, accordingly, for a general scarcity of vegetation and fauna. The dry layer of

Above: An evocative panorama of the rainforest in Costa Rica.

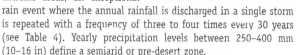

4

Place	Date	Average Annual Precipitation (mm) [inches]	Maximum Precipitation per Single Rain Event (mm) [inches]
Chicama (Peru)	1925	4 [0.2]	394 [15.4]
Auzou (central Sahara)	May 1934	30 [1.2]	370 [14.4]/3 days
Swakopmund (Namib)	1934	15 [0.6]	50 [1.9]
Lima (Peru)	1925	46 [1.8]	1,524 [59.4]
Sharjah (Trucial Coast)	1957	107 [4.2]	74 [2.9]/54 minutes
Tamanrasset (central Sahara)	Sept. 1950	27 [1.1]	44 [1.7]/3 hours
Bisra (Algeria)	Sept. 1969	148 [5.8]	210 [8.2]/2 days
El Djem (Tunisia)	Sept. 1969	275 [10.7]	319 [12.4]/3 days
Themed (Sinai)	1972	30 [1.2]	142 [5.5]/1 day
Djanet (Sahara)	1976	19 [0.75]	42.4 [1.7]/1 day

Right: An enormous iceberg off
Antarctica.

Below: Uadi Rum in the Jordanian
Desert; below right: absence of soil
in Egypt's White Desert.

air typically found over arid regions promotes rapid evaporation and exacerbates soil dryness, causing drastic thermal variations and accelerates erosion that shapes the landscape. In an environment with a humid climate, the ground absorbs 40% of the daily solar insolation, while water and plant cover absorb 30%. Another 20% is reflected by clouds and 10% by dust particles. In humid climates at night, the ground reradiates 50% of the heat accumulated during the day back toward the sky; 20% is reflected by clouds, 10% by dust, and 20% by the vegetation cover.

In arid zones, however, 90% of daytime solar radiation heats the desert soil and the lowest layers of the atmosphere; only 10% is reflected by dust suspended over the ground and by the cloud cover, if present. At night, nearly 100% of the accumulated heat is released from the soil and dispersed. Of this, 90% rises into the sky while 10% is sent back toward the usual low-lying haze. The drastic drops in temperature from levels above 50°C (122°F) on torrid summer days to less than –25°C (–13°F) at night are an important regulatory factor at the biological level for the survival of flora and fauna.

CLIMATIC CONDITIONS THAT CREATE DESERTS

A desert is a particularly arid region of the planet where the vegetation cover is limited or almost absent and the fauna is scarce. It is created by the convergence of certain climatic conditions with particular factors—in many ways extreme—that surround the area.

Left: Vegetation is rather scarce throughout the Western Egyptian Desert (the White Desert). Below: silver spears of the haleakala, a shrub indigenous to the Hawaiian Island of Maui

Geographic position and the distance from the sea and other humid zones define the **continentality** of a region. Every zone west of subtropical regions is thus delineated and governed by moisture-free winds. The dynamics of cyclonic **atmospheric circulation** are also important, as they are responsible for precipitation itself and for a territory's orography, the branch of physical geography dealing with mountains.

Temperature is another distinguishing characteristic of deserts. Monthly averages are often extreme with highs hitting 60°–70°C (140°–158°F) in hot deserts and lows around –30°C (–22°F) in cold

5		
Hectares	Acres (millions)	
Surface area of Earth	51,800	128,000
Earth's land area	14,750	36,448	(28%)
Land under cultivation	1,450	3,583	(2,8%)
Arid land	4,900	12,108	(9,5%)

Left: Table 5 shows the extent of arid and cultivated areas as a percentage of Earth's land mass.

ones. The difference in temperature between the ground and the overlying air provokes turbulent movements of the irregular winds that act as erosive and transporting agents. The aridity of a region can also be influenced by the **albedo**, which is the difference between the solar irradiation that strikes a surface and the radiation

18

Top: Satellite views of cloud formations; above: Partially eroded stratified rocks in the Negev Desert in Israel.

reflected into the atmosphere from that same surface. Other factors that contribute to the formation of a desert are its **tectonic-structural makeup** and the **lithology** of its rock substrate.

Empirical formulas are used to distinguish an arid from a semiarid environment; one of these defines the **aridity index (I)** derived from the ratio between precipitation and evapotranspiration calculated on an annual basis (see Table 6). Another index of aridity considers the average annual precipitation (P) and temperature (T). The formula is expressed as I equals P over T + 10, or (I = P / T+10). An "I" value of 10 or less defines an arid climate; deserts have "I" values of 5 or less, and semidesert zones have "I" values between 5 and 10.

$$I = \frac{P}{T + 10}$$

P/ETP (P precipitation; EPT evapotranspiration)	
subhumid zone	$(0.50 < P/EPT < 0.75)$
semiarid zone	$(0.20 < P/ETP < 0.50$
arid zone	$(0.03 < P/ETP < 0.20)$
hyperarid zone	$(P/ETP < 0.03$)

THE PLANET'S DESERT ZONES

A UNESCO study based on all the parameters, values and characteristics described above concludes that desert and semidesert zones exist in at least 53 countries or areas on our planet (Table 7). Excluding the polar zones, which would merit a separate treatment, deserts are distributed across the band of territory between latitudes 15° and 40° North and South. Depending upon their geographic position

A beautiful example of a natural arch in the Jordanian Desert's Uadi Rum.

and other characteristics—especially temperature—they subdivided into various types. We arrive at three main categories, each containing more precise distinctions: **hot**, **transitional** and **cold deserts** (or better, those with a cold winter season).

Extremely arid continental deserts fall within the first category, where frost has no influence whatsoever. These are character-

7

Afghanistan	India	Peru
Angola	Iran	Puerto Rico
Argentina	Iraq	Russia
Australia	Israel	Saudi Arabia
Botswana	Italy	Somalia
Bolivia	Jordan	Spain
Brazil	Kenya	South Africa
Cameroon	Lebanon	Sudan
Canada	Libya	Syria
Chile	Madagascar	Tanzania
China	Mexico	Tunisia
Dominican	Mongolia	Turkey
Republic	Morocco	United States
Egypt	Mozambique	of America
Equatorial Africa	Nigeria	Venezuela
Ethiopia	North Africa	Yemen
Greece	Oman	Zambia
Haiti	Pakistan	

ized by scarce and irregular precipitation, sometimes less than 120 mm (4.7 in) a year. Years may pass with no precipitation at all, then an entire year's average may fall in a single event. Far lower averages were recently documented in Algeria's In Salha region, where less than 15 mm (0.6 in) of average annual rainfall was measured during 15 years of observation. Other distinctive features of these extremely dry areas are lack of moisture on the ground and in the lower atmosphere, and evaporation levels at 4,000 mm (156 in) with evapotranspiration (the combined effect of loss of water from the ground directly to the air and the "exhalation," usually at night, of moisture by plants) at

Rocky formations typical of the Syrian Desert.

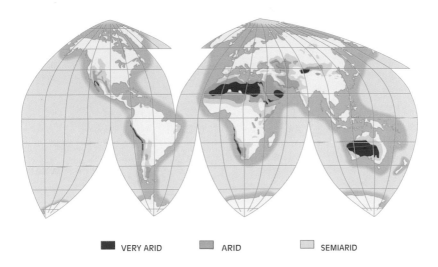

■ VERY ARID   ■ ARID   ☐ SEMIARID

2,000 mm (78 in) representing a quantity of water 20 times greater than a whole year's rainfall. Temperatures are high throughout the year and thermal fluctuations are extreme. Algeria offers an example: variations between day and night temperatures of up to 37.6°C (99°F) have been measured there. Deserts in this category are the Kalahari, Namib, part of Australia's Great Sandy Desert, the Sonoran Desert between Mexico and the United States, and a few areas in Iran, Pakistan and Arabia.

Extremely arid coastal continental deserts are found where subtropical ocean air masses (cells) characterized by high pressure meet cells of high-pressure continental air. These two systems carry no moisture, despite the ocean's proximity, and their convergence gives rise to dry, stable air masses that extend onto the ocean and toward the interior. The rains that fall with pronounced variability measure less than 250 mm (10 in) a year, even though frequent cloudiness would suggest higher numbers. Humidity generally persists above 50% with significant amounts of dew measuring about 50 mm (2 in) per year. Fog occurs often in these areas. Fog banks form in the lower level of the atmosphere that stretches across the cold ocean waters. Average annual

Above: This map is taken from documents furnished by UNESCO and shows the global distribution of extremely arid, arid, and semiarid zones.

Below: An area of evaporation with obvious saline formations in the Atacama Desert in Chile.

21

Above: Ripples on the sandy surface of the Libyan Desert; at right: cactus in Mexico's Vizcaino Desert, Baja California.

temperatures fall around 18°C (64°F), cooler than those in continental deserts by an average 5°C (41°F). This variance is caused by flow toward the equator of cold ocean currents such as the Benguela and the Humboldt west of South America. Thermal ranges are rather limited (average annuals of about 6°C (43°F) have been recorded), but evaporation measures 850 mm (33 in). Winds blow with considerable force and constancy and always in a single direction (unidirectional). Among the most important morphogenic agents, they are responsible for creating a strange and most dramatic landscape (discussed in depth on page 62). Abundant salts are suspended in the lower atmosphere; their crystallization and growth on rocks is a potent weathering mechanism.

The deserts in this category are the Atacama in Chile, the deserts of Baja California in Mexico, the Namib Desert on the southwestern coast of Africa and the coastal deserts of Australia and Morocco.

Arid continental deserts in the middle latitudes and **mountain deserts** at high altitudes owe their existence to complicated atmospheric circulation. Situated between latitudes 30° and 50°, continental air masses assemble during the summer season in these territories, but the same areas are governed in winter by the flow of polar air masses that coincide with the Canadian and Siberian anticyclones, which are moderately moisture-free. In other cases, such deserts are caused by masses of humid ocean air that rise against mountains encircling desert areas.

The rising air mass cools and loses its capacity to hold so much water, resulting in heavy precipitation on the windward side of the mountains. The now-waterless air mass moves downslope, becoming warmer and regaining its ability to hold the moisture that is no longer present. The arid zone on the leeward side of the mountains is called a "rain shadow."

Left: A true "cathedral in the desert," this lofty butte stands tall in Monument Valley in the United States.

The imposing Himalayan chain illustrates this mechanism in action as it blocks the mass of rain-bearing air that would otherwise arrive from the Indian Ocean. The added barrier of constant western winds also keeps moisture-charged masses of maritime air from reaching these regions. Relative humidity varies between 30% and 40%, and evaporation removes 1,500–2,000 mm (59–78 in) of water every year. Temperatures average less than 18°C (65°F), and the wind fluctuates from medium to strong. Another example of a rain shadow occurs in Washington State. Air rising over the Olympic Peninsula loses its moisture there, making Olympic National Park a rainforest. The drier central and eastern parts of the state are in the rain shadow.

The rainfall in arid continental deserts puts them on the threshold of regions considered transitional, experiencing highly variable rainfalls of 250–500 mm (10–20 in) per year, concentrated largely in spring and summer, with minimal moisture. Their winter season is rather long, even if the impact of freezing on the environment and landscape is limited to higher elevations. Examples of this desert type are found in North America, Asia and the extreme southern regions of South America.

Below: Diagram of the fog-forming mechanism in Chile's Atacama Desert. Fogs are induced by subsident air over a coastal thermal inversion.

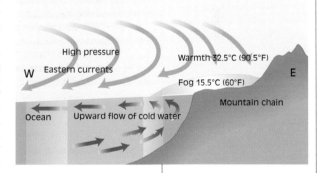

Cold deserts. Precipitation varies wildly in these regions, especially between the end of spring and the beginning of summer, and is largely regulated by the topography of the territory itself. Snow-

Top: A view of the Gobi Desert in Mongolia.

Below: The Atacama Desert in Chile.

fall becomes significant at high elevations. Despite the fact that the cloud cover is thick and continuous compared to other desert types (such as those at low latitudes), the terrain is insolated (exposed to the sun) only to 60–70% of possible levels. Annual temperatures average around 10°C (50°F), with maximums reached on summer days, and average below freezing during at least two winter months; nights are frigid all year long at high altitudes. Given these rather reduced temperatures, evaporation does not exceed 500 mm (20 in). Mongolia's Gobi is an example of a cold desert.

Polar ice cap deserts. The polar zones north of the Arctic Circle and south of the Antarctic Circle may be anomalous in the context of this book, yet they truly merit a place in the inventory of deserts. Their morphologies are quite different, even if the scarcity or absence of soil is a common factor and their morphogenetic and pedogenetic agents are ultimately different or behave differently.

Climatologist W. Koppen defines polar deserts as areas perpetually subjected to freezing weather whose average temperatures never really go above 0°C (32°F) and where the world's lowest temperatures are recorded. A record low of –88.3°C (–126.9°F) was measured at the end of Antarctica's long polar night near the Russian meteorological sta-

tion Vostok situated on a plateau 1,300 km (800 mi) from the South Pole. Average temperatures around –60°C (–76°F) were recorded for the months of July, August and September at the Amundsen-Scott station, while the average near the American McMurdo station is –32°C (–26°F). McMurdo enjoys the ocean's mitigating influence and sits at a lower altitude.

Temperatures are slightly higher at the other pole. Annual averages are about –23°C (–9°F) on the Arctic surface, which is relatively low-lying. Temperatures average between –30°C to –35°C (–22°F to –31°F) on Greenland's ice cover, where February is the coldest month with lows near –50°C (–58°F). July is the warmest at –11°C (–12.2°F), so the annual fluctuation is 39 Celsius degrees. Violent winds (**blizzards**) carry huge quantities of snow crystals that accumulate on smooth compacted surfaces that stretch for miles. Precipitation is not rare and occurs in the form of snowfalls.

Below: Ayers Rock in Australia, better known as Uluru in the Aborigines' language, has always been this continent's "sacred place."

WEATHERING

Overleaf: *The Olga Mountains in Australia.*

Above: *A majestic sandstone formation in Monument Valley, U.S.A.*

Deserts are characterized by two kinds of surface topography, **deserts on shields** or platforms, and basin deserts.

The first are typical of exceptionally stable seismic zones (stable tectonics) such as the plains of Africa (the Sahara and southern Africa), most of the Saudi Arabian peninsula, and parts of Central Asia, India and Australia.

Shield deserts are dominated by erosional surfaces that are often established on rocks of volcanic origin found at the base of a stratigraphic sequence, as in western Australia and southern Africa, or on wide shallow basins like the drainage area of Lake Eyre in Australia or Lake Chad in Africa. A third platform consists of surfaces that developed on horizontally layered sedimentary rocks, like the plateaus of Colorado and central Asia, or on the outcrop of Nubian sandstone in North Africa. The origin of these plains is not connected to their current aridity. Their topographic features include deep canyons in the plateau region of Colorado or in northern Namibia and only locally elevated promontories and isolated hills called **inselbergs**, formed by differential erosion.

Intra-mountain basin deserts are dominated by a succession of mountains and troughs often characterized by endorheic (closed) drainage systems. These are common in active seismic areas such as the southwestern United States; the arid littorals of Chile and Peru; and Iran, Afghanistan, Pakistan and a portion of central Asia.

Special and sometimes unique morphologies are found within

these various environments. They may occupy small areas or limited zones but are always, as mentioned earlier, invariably linked to three basic elements: climate, geology and the age of the surface itself.

Desert surfaces are subject to weathering and mass wasting processes. The most frequent are atmospheric, mechanical (physical), and chemical in origin, and are described more broadly in the following paragraphs.

The shapes they engender are found in many deserts, however different and far from one another they may be. Similar morphologies do not guarantee, however, that the same geological mechanism generated them. On the ground, different processes can give rise to comparable forms in different places, starting from different beginnings—akin to convergent evolution in biology.

Below: Examples of differential erosion in the shaping of rocks eroded by atmospheric action in Egypt's White Desert (Western Desert) and, at bottom, in the Jordanian Desert.

METEOROLOGICAL WEATHERING

Weathering affects both rock outcroppings and the detritus made up of sand or cobbles, and is characterized by being **surficial** and **selective**.

The surficial nature of this weathering is because water infiltration and temperature variations do not penetrate very deeply, whereas weathering in humid climates affects a considerable thickness of rock.

The selectivity of atmospheric weathering derives instead from the fact that desert rocks are exposed rather than covered by soil or vegetation, so lithologic and structural characteristics of the rocks control their weathering. Thus, fractures, faults, crystal edges and other planes of weakness exposed at a rock's surface tend to direct, concentrate and control the process of its degradation.

Finally, an alteration varies according to the rocks' exposure, which depends in turn upon local variations of temperature and humidity.

Surface rocks are subject to phenomena of splintering, desquamation (or exfoliation), breakage and granular disaggregation. The final result is the creation of a huge quantity of debris.

Splintering (shattering into coarse, angular flakes) and breakage (pebbles and spherical blocks that look like they were split by an ax) happen in various kinds of rock, but rarely in rough-grained igneous rock. Split rocks have been reported in many deserts, and salt crystals are often observed along the fracture planes.

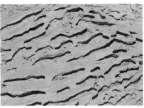

Above: Examples of atmospheric weathering, which is caused by rain, wind or other forces of the atmosphere, on rocks of different composition; the above rock was probably shattered by seesawing temperatures.

Desquamation, or the formation of flakes parallel to the surface, can occur on isolated boulders, where it is called "onion peeling," or across entire outcroppings of rock. To occur at all, the material must expand somewhat in response to fluctuating temperatures. Second, foreign substances such as ice, salt or roots that are likely to be present must expand, and finally, one of several chemical changes must occur (water absorption by colloidal substances, hydration, or oxidation of the silicates). Pre-existing discontinuities—caused when the rock itself was cooling off during its formation—can lead to water infiltration and, in turn, give rise to a chemical change. This flaking process is probably the main cause of natural arches in sandstone, which are created when rocks are laminated horizontally with vertical fractures and a few of their horizons are less resistant. Significant examples can be observed in southern Utah in the United States.

Granular disintegration, or rocks breaking into single crystals, occurs normally in any polycrystalline rock, including coarse-textured igneous rock and metamorphic rock. It is caused by both chemical and physical processes but always involves rocks in which at least some minerals have been chemically altered.

Weathering can produce peculiar features, not necessarily exclusive to deserts. They include **tafoni**, the result of **alveolar weathering** or pitting, **perforations** and pits.

TAFONI

Tafoni are cave-like formations that are usually several cubic meters in volume and feature an arch-like entrance. Their concave internal walls are generally swollen and friable, easily fractured, and may be suffused with salt. Their floors are smooth, nearly horizontal and covered by detritus.

Tafoni generally appear in groups, having transformed the host rock into a kind of Gruyère cheese. The surfaces of the caverns and the air inside them are often colder and more humid with temperatures and relative humidity that fluctuate less than those outside—conditions that facilitate weathering. Tafoni are found in various climatic conditions in coastal environments like the Gallura in Sardinia and in many deserts, including the Mojave, Sonora, Antarctic, Atacama, Sahara, Saudi Arabian and in southern Australia. They develop in many varieties of rock and have been observed in argillite (schist or slate), porphyry, rhyolite, limestone, sandstone, gneiss, and granite, where the most spectacular forms are found.

The origin of tafoni has always been a subject of controversy and debate; salt weathering is most often called to account. Certain flared shapes found in the lower sections of steep walls on granite inselbergs are linked to tafoni.

A few tafoni and alveolar formations in Israel's Negev Desert, where the surface is heavily eroded by chemical and atmospheric action.

Below: various natural arches in Algeria, Jordan and Libya.

ALVEOLI

Alveoli are small hollows 5–50 cm (2–20 in) in diameter that form honeycombed structures when they appear in groups. Morphologically, they are similar to small tafoni and both

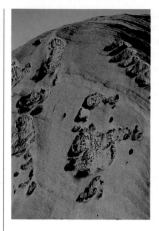

Above: A few of the most widely found configurations of honey-combed alveoli.

Below right: Weathering pits in the Western Egyptian Desert.

forms are sometimes present, but no common origin is implied. Alveoli are found in various climates from deserts to coastlines and on various types of rock. Like tafoni, alveoli have controversial origins but there is a general concurrence on the importance of salt weathering. Thin partitions that can be relatively resistant to atmospheric weathering separate the individual alveolar pits.

Weather pits are not uncommon desert shapes. Frequently found on bare, horizontal, or slightly inclined granite, sandstone, porphyry and other thick homogeneous rocks, they may be as wide as 15 m (50 ft) and 4 m (13 ft) deep. In plan, and depending upon structural factors, their shapes are highly variable but often elliptical or circular on horizontal surfaces, and asymmetrical on sloping surfaces. In transverse section, they usually have a flat, semicircular base with bulging walls.

Studies conducted on Australian granites indicate that these pits take shape through breakage concentrated at points of physical weakness or along fractures, through surface flaking of the granite, or through disintegration by lichens. Once

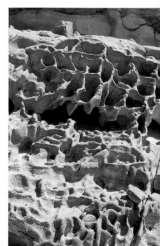

they are in place, water stagnates in the pits and, as a consequence, chemical weathering becomes significant. The bulging shape of the pit walls may occur because water sits on the lower part of the pit longer and continues to alter the rock by penetration. Alteration by salt at the surface of a terrain near a capillary fringe is the probable cause of structures called **zeugen** or **pedestal rocks**, which were traditionally attributed exclusively to wind abrasion.

MECHANICAL WEATHERING

The presence of acutely angular broken rocks and detritus in the vast majority of deserts spawned the idea that mechanical (or physical) weathering may be more important than chemical weathering. Clearly, mechanical processes are more important than chemical or biological ones in arid and semiarid regions, but the latter cannot be undervalued since humidity is relatively abundant in deserts, especially in the form of dew, and relative humidity is often high at night. Moreover, some mechanical operations such as salt weathering actually depend on chemical agency.

THERMOCLASTISM

The concept of **insolation weathering**, or thermoclastism, suggests that rocks and minerals break up as a result of wide-ranging daily and/or seasonal temperatures as well as the big contrast between the heat on a rock's surface and at its core. These fluctuations cause rocks and minerals to expand and contract at different rates, depending upon their coefficients of thermal expansion and other properties of a material that may induce the breakage.

This notion is fairly deep-rooted and is vindicated by the obvious and intense thermal fluctuations in deserts. It is also sustained by accounts of sounds like rifle shots being heard in the desert at night; no one, however, has ever been on hand to watch the rocks fracture.

Above: Pedestal and mushroom rocks in Libya, Egypt, Libya and Israel, respectively

Most studies and laboratory tests have demonstrated, in fact, that temperature variations alone are not enough to fracture rocks. Weathering by dry insolation, however, can induce micro-fractures that increase the rock's permeability and open the way to other erosive processes. Where there is humidity associated with wide-ranging daily temperature, water trapped in the capillaries of the rock can generate internal pressure great enough to fracture it. This process

Right: A natural arch in the Eastern Egyptian Desert.

Below: A rounded rock in Egypt; at bottom: rocks shattered by arctic frost action in Norway.

of weathering by humid insolation depends upon the structure of the pores of the material, thermal variation and the availability of water.

If solar insolation fails to produce immediate effects, however, it is possible that repeated heating and cooling over long periods of time will also provoke the detachment of rock splinters.

Variations in temperature and humidity occur simultaneously in nature. Variations in humidity cause chemical reactions with minerals, and increased porosity leads to increased water and air, which are better conductors of heat than the rock.

CRYOCLASTISM

This term describes how rock crumbles when water freezes in the cramped spaces inside a rock, and the expanding volume of water crystals exercises pressure. The phenomenon has not been studied very much in deserts, but it is an active force in high-altitude deserts where temperatures regularly drop below freezing. It occurs in Africa's Atlas Mountains, the Eurasian deserts, and in most of the Great Basin and the mountain chains in Iran, Afghanistan and Mongolia. Moreover, it was clearly active in the past—during the coldest and most humid periods of the Quaternary—and consequently, some of the rubble now found in arid zones was probably created by cryoclastism in remote times. Another factor is the presence of salts in solution found along the fractures in rocks. This is a common occurrence in deserts and it causes the ice's fusion temperature to vary, which delays cryoclastism (if there isn't much salt) or augments it (if the salts are especially abundant).

34

WETTING AND DESICCATION

Repeated cycles of water absorption (wetting) and desiccation (drying) lead to the disintegration of certain types of rock such as those in the clay family or fine-grained rocks in general.

The desert surface can be wetted by **rain**, **fog** or **dew**. Dew can be very significant in certain deserts by offering a daily cycle of wetting and drying. Jordan experiences more than 50 dew days each year, and more than 20 mm (0.8 in) per year of precipitation from dew has been documented in some places. In Israel, wetting by dew will penetrate a few millimeters below surfaces covered by vegetation. Since the dew in Israel is salty, weathering is magnified.

HALOCLASTISM

Haloclastism, or salt weathering, is one of the most active processes in hot and cold deserts but is rare in temperate zones.

The most favorable climatic conditions for salt weathering are those typical of hot arid environments: low relative humidity and heavy evaporation during the day, high diurnal temperatures and fluctuating relative humidity. Under these conditions, saline solutions are subject to evaporation and to temperature variations that favor the growth of crystals and provoke a volumetric change in salts with higher coefficients of thermal expansion. In addition, strong deviations in humidity provoke equally strong changes in vapor pressure that are favorable to salt hydration.

Evaporation from the top desert strata induces the continuous capillary ascent of water above the **piezometric surface** of groundwater. The capillary fringe, or the area immediately above the water table where capillary action occurs, becomes saturated with salt due to this continual evaporation.

The thickness of this fringe varies, especially as a function of the granulometry, or particle size, of its sediments. In the case of very fine sediments and extreme desert conditions, it may be more than 3 m (10 ft) thick. The fringe may lie immediately below the desert surface where its upper edge can host a growth of crystals, which are often made of gypsum and produce the so-called **desert roses**. In other cases, it reaches the surface by creating an area of salt blooms and crusts, often in the form of **needle-shaped crystals** that spurt out of the ground.

Above: Two examples of sandstone and limestone partially and totally eroded in Egypt.

Below: Saline formations emerge on the surface after evaporation.

Above: Delicately embroidered saline horizons and salt concretions in Jordan's Dead Sea.

A common source of the salts found on desert surfaces is the deposition of aerosols (colloidal substances in suspension) that are primarily transported from the oceans. In this scenario, the concentration of salts is reduced as they travel inland from the coast. Other sources include dry bodies of water such as lakes and lagoons, dusts and volcanic gases, and deposits of fossil salt.

Salts also derive from phreatic waters, which acquire their salt from both rain water and buried salt deposits, from the sea water that penetrates the subsoil of coastal plains (often due to the prodigal extraction of fresh water) and, perhaps most important, from the chemical alteration of rocks. The many kinds of salt vary from place to place. In Australia, for example, sodium chloride (NaCl) predominates, and Chile is associated with calcium sulfates and sodium, while bicarbonate of soda (NaHCO$_3$) is widespread in the salt lakes of eastern Africa. The type of salt is important because it determines the very nature of the alteration.

Normally, salts are scattered and only found in small quantities. The average concentration of nitrates in the desert soils of Chile, the Sahara and the Mojave, for example, is in the order of 0.1–0.2%. Chilean nitrates are distinguished from those in other areas by their unusual concentration in certain locations and occasional separation from other salts. These mechanisms of concentration and separation are fundamental to the weathering process.

Salts are concentrated in three main environments: **humid deserts** on arid western coasts, **playas** and **sabkhas**.

In humid deserts on arid western coasts like the deserts of Chile, Peru, Namibia and western Australia, ocean salt can accumulate very quickly. Marine salt can accumulate inland, too; in the Thar Desert, the annual quota of marine salt is estimated to exceed 200,000 tons.

Salt also accumulates within and around closed (endorheic) drainage basins including **playas**, **salt lakes**, **chotts**, **pans** and **kavirs**. Playas may contain fossil salts transported by the sheetwash of surface waters, by wind action, or else by the emergence of groundwater. **Sabkhas** may form in areas of coastal aggradation, especially along inland seas like the Arabian Gulf and the Red Sea. The capillary ascension of sea water is the primary cause of extensive accumulations of salt on these wide **intertidal plains**. In basins, salts precipitate on the bottom in concentric rings. Chlorides, which are the most soluble salts, are in the center followed by an intermediate zone of sulfates and an outer ring of carbonates, which are the least soluble. Such arrangements are found in California's Death Valley and in the Bonneville Salt Flats in Utah. Chile's saline depressions also exhibit crusts of rock salt girdled by a ring of sulfates. A comparable arrangement is observed in a vertical section of ground. Salt concentrations diminish with depth, and the sulfates deposit themselves on a higher level than the more mobile and soluble chlorides.

The presence of nitrates on slopes that are relatively distant from the floor of the Atacama Desert's central depressions is due to the capillary migration of residual solutions from the saturation zone toward the top of the slope. The vertical and horizontal distribution of salts reflects the mobility of negative ions in solution. Sulfate ions migrate more slowly and are concentrated near the surface near the saturated zone; nitrates occupy an intermediate position, and the more soluble chlorides migrate farthest and fastest.

The type of salt or mixture of salts in attendance is important in haloclastism. The most common are calcium carbonates, especially in semiarid regions, various calcium sulfates in arid regions, and the sodium chlorides chiefly found in coastal areas. But nitrates, carbonates, chlorides and sulfates are commonly found, too, especially of sodium, magnesium and potassium. The effectiveness of salt weathering varies with the chemical properties of salts such as their solubility, crystallization and hydration. The most effective are the sulfates of sodium and magnesium. In terms of sodium sulfate, for example, the hydra-

Below: An impressive accumulation of salt near the Dead Sea in Jordan.

37

tion of thenardite (Na_2SO_4) in mirabolite (Na_2SO_4 x $10H_2O$) causes a considerable change of volume; its solubility depends on temperature variations, and it is so soluble that it is always abundantly available. Crystals grow when saturated solutions evaporate, and the crystals tend to have elongate or needle-like forms, which render them more effective in breaking pores.

Weathering by salt is also significant in a practical context in that it damages buildings, roads and other engineered products in deserts. The control and management of salt weathering depend rigorously upon knowing the composition and depth of phreatic waters, the depth of the capillary zone, and the relationship among these parameters and the depth and composition of the substructure.

This page: Panoramic images of the Dead Sea in Jordan.

Various studies conducted in the Middle East suggest that the risk of weathering by salt is directly proportional to the proximity and salinity of the water table, and can thus be predicted through a spatial analysis of these variables. The disaggregation of rock by salts therefore depends on chemical and mechanical processes.

Chemical processes have been analyzed in detail, especially in relation to the manufacture of cements and reinforced concrete, but are still underrated in natural desert environments. The three main physical processes can be summarized as the growth of salt crystals in confined spaces by evaporation and/or the cooling of saline solutions, the growth in volume of salt crystals brought about by hydration, and the thermal expansion of salt crystals due to insolation.

CRYSTAL GROWTH

According to some authors, the growth of crystals from saturated solutions onward is the primary haloclastic process. Crystal growth depends upon diminished solubility, which is proportional in turn to declining temperatures, the evaporation of a solution, or a mixture of different salts in solution.

HYDRATION

Many salts are found in both their anhydrous and hydrated states and, in this case, can expand by absorbing water into the meshwork of their crystals. The hydration of calcium sulfate ($CaSO_4$) increases its volume. Thus, desiccated or even partially dehydrated salts can exercise pressure on the walls of the fissures and pores that contain them once they are wetted. The various forms of anhydrous and hydrated $CaSO_4$ are widely distributed in deserts. The supposed mechanisms first involve the crystallization of a hydrated form in a fissure or pore of the rock, followed by its slow dehydration at the hand of high desert temperatures. Later, an abrupt wetting causes rapid hydration, followed by strains against the walls of the fissures.

Below: Commonly known as "desert roses," these rocks are the result of gypsum crystal growth and often assume complex and spectacular forms.

A daily cycle proposed for Chilean deserts suggests that saline solutions rise to the surface by capillary action during the day and crystallize in their anhydrous phase. When the temperature falls at night, low-lying anhydrous and hydrated salts absorb atmospheric water vapor and expand.

Disaggregation occurs fastest when the soil or atmosphere provides water in moderate quantities. Sodium sulfate is the most effective salt, while sodium chloride is at the other end of the scale because it crystallizes slowly. The rocks most likely to crumble are the most porous ones. Large-grained granites are among the most susceptible while fine-grained, less porous rocks are not responsive to the activity of salts.

THERMAL EXPANSION

If heated, salt crystals can expand according to their thermal characteristics and the fluctuations in temperature to which they are subjected. Many desert salts have a high coefficient of volumetric expansion. Temperature changes can exert a considerable impact on salts precipitated within fissures and pores close to the surface of rocks. According to many authors, however, laboratory experiments indicate that thermal expansion in the absence of moisture is almost inconsequential.

WEATHERING BY ALGAE AND LICHENS

A large variety of algae and lichens are common in deserts. In the Atacama Desert, 150 species of lichen have been collected; in the Sahara, more than 130; and in the Negev, more than 50. They are localized in surface rubble, in the subsoil, on the underside of partially buried stones, above and inside rocks, and are found in both hot and cold deserts.

Algae and lichens can be distinguished as **endolithic** and **epilithic** organisms. They first live inside the pores and along the fissures of light-colored rocks (marble, granite and granodiorite) up to a depth of 40 mm (1.6 in). The latter are typically blue-green algae (*Glaeocapsa*) and lichen such as *Caloplaca alociza*. On the contrary, epilithic organisms such as the well-known and widespread crusting lichens attach firmly to rock surfaces using thalluses. The consequences include a modified micromorphology of the rock's surface, granular breakdown, and the desquamation or even outright dissolution of the rock itself.

This page: The erosion and weathering of limestone surfaces characteristic of the White Desert in Egypt.

LIMESTONE WEATHERING

In deserts, the alteration of limestone, commonly called **karstic weathering**, can be charged in part to organisms, as seen earlier, and partially to dissolution by acid rain. Such forms are frequent despite the effective absence of water and vegetation. Displays of karstic dissolution such as sinkholes, pits, crevasses and karstic corridors, caves, dolines, flutes and furrows (known by the German term **rillenkarren**) have been observed in Australia, Namibia and in the Sahara. When analyzing any desert karst forms, it is essential to remember that many of these shapes may have been inherited from earlier humid climates.

ROCK PATINAS

The phenomenon of mineral coatings on rocks is not limited to deserts but also occurs in high mountains, near the poles, along the coasts, and near river channels under a variety of climatic conditions. All the same, patinas are undeniably thicker and more extensive on denuded desert rocks than anywhere else.

Rock varnish, which is rather incorrectly but commonly called desert varnish, is the most diffuse and dominant weathering that manifests in the form of coatings, so much so that in U.S.

deserts it covers 75% of the denuded rocks. It is a hard crust from 5 to about 100 μm thick on the surface of the ground, pebbles and rocks. Its thickness varies enormously according to the ruggedness of the surface, its exposure to the sun and to wind abrasion. Its color is usually black or dark red according to its manganese content.

Left: A desert patina on a rocky sandstone surface.

Below: Differential erosion and weathering patina on limestone rock in the Jordanian desert.

The varnish is hard, no matter what kind of rock it covers. Indeed, the underlying rock can be weathered and can crumble completely while the varnish endures and preserves the original shape of the rock. Early on, scholars thought that the constituent elements of varnish must have come from the rocks they covered, but it has been discovered that these elements, including iron and manganese, come from wind-entrained dust. Various microorganisms such as lichens and bacteria are needed to attach the manganese; they grow on particles of clay and acquire energy from the oxidation of manganese during the short humid periods when they become active. Manganese absorbed into the clay minerals stays put and is retained among the internal spaces of the crystalline mesh. Varnish thus protects bacteria from the dry conditions of the desert surface.

Once started, the manufacture of patina is a process that feeds itself and further entraps clay and manganese. Patinas do not form if the environment is reducing (has a high **pH**) or when the organisms that bind the manganese are supplanted by mosses, lichens or higher plants (see page 40). Patina growth is inhibited if rock surfaces are too smooth, if a rain-wash or dehydration is too intense, or if there is too little manganese available. They are broken or dismantled when either humidity or wind erosion increases.

The formation of a varnish in just 13 years was observed in Israel,

41

Clockwise: A weathering patina on faceted rocks in the Libyan Desert; nodules of manganese and desert varnish in Egypt; the Uadi Rum in Jordan.

Facing page: The floor of a playa lake in Libya, and a small water well in the Algerian desert

yet it is generally thought that varnishes, and especially the thicker ones, take far longer to form.

A technique for dating varnishes was recently formulated based on the variations over time in the proportions of certain cations (titanium, potassium and calcium). This technology has established minimum ages of 5,000 years for certain varnishes in Australia and more than 10,000 years in some U.S. deserts. Using other dating methods, ages over 30,000 years in Utah and a million years in Nevada have been ascertained.

Desert glass, or silicon glaze, is another type of patina. It is a shiny, colorless covering, rich in silicates, with an almost ceramic aspect and about 0.01 mm (0.0004 in) thick. It is found on some cobbles, boulders or other materials rich in silica such as flint, agate, fossil wood or quartzite. Desert glass seems to be very common and is probably linked to the dissolution of silica and evaporation.

One final phenomenon is **case hardening**, which involves the formation of a coating from 0.5–5 mm (0.02–0.2 in) thick, much clearer and harder than the underlying rock from which it is separated by a distinct boundary. It is made up of a calcareous surface horizon and an underlying kaolin-enriched zone. Case hardening is not limited to desert surroundings even though, like rock varnish, it develops more frequently there.

PLAYAS AND SABKHAS: THE GREAT PLAINS AT BASE LEVEL

*The areas at lower elevations in closed (endoreheic) desert drainage basins often feature nearly horizontal surfaces that are free of vegetation and composed of fine sediments. Due to their prevalence and characteristics that are quite different from their surroundings, these base-level plains assume a variety of local names: **nor** (Mongolia), **pan** (South Africa), **sabkha** (North Africa and Middle East), and dry lakes and **playas** (North America).*

*The variety of local terminology is evident in the Sahara, where such plains are called **sebkh** (**sebjet, sebchet, sebkha, sabkhah,** and **sebcha**), **zahrez, chott** and **garaet.***

*"Playa" is the best-known term, but different types of playas have acquired specific names. For example, a clay playa is known as **clay pan** in Australia, **takyr** in Russia, **khabra** in Saudi Arabia and **qu** in Jordan; and a salt playa is called a **kavir** in Iran, **salar** in Chile, and **taska** in Mongolia.*

*Coastal plains, known as **sabkha** in Arabic, are very similar to inland playas. These are nearly horizontal surfaces with fine-grained sediments and free of vegetation.*

The playas are important locales for mineral concentration and are valuable indicators of climatic changes. More than 1,000 playas are known in North Africa and about 300 in the United States, of which at least 120 are the remains of Pleistocene lakes.

The existence of playas is primarily controlled by climate but, given that many are located in endoreheic basins,

other causes compete to determine their formative conditions. Endoreheic basins are generally the result of tectonic movements such as faults (California's Death Valley), folds or subsidence (Chott Djerid in Tunisia), but also eolian or alluvial deposition (interdune playas in the Empty Quarter of Saudi Arabia), erosion (the Ras Qattara depression in Egypt), or volcanic activity (the Puna in northern Chile). The extent of a playa is a function of the size of the endoreheic

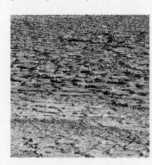

basin and varies by a few square meters to more than 9,000 sq km (3,500 sq mi) in Australia's Lake Eyre. The nature of playas depends on many interrelated variables, the most important of which are groundwater, runnels, surface water, interstitial water, eolic processes, and chemical and biological reactions.

Playas are characterized by fine-grained sediments derived from eolian activity, surface rivulets and salt deposits. The accumulation of fine-grained sediments has three principal geomorphologic results:

- *playas tend to be relatively impermeable;*

- *the phenomena of desiccation are common;*
- *eolic activity is intense.*

Water on the playas comes from direct precipitation or groundwater and is dissipated by evaporation. Playas exist only in areas where annual evaporation is significantly greater than annual precipitation. The moment this ratio is reduced and water content increases, however, the water's ability to stay on the surface is extended and lakes

can eventually form. The duration of high water is variable from just one season to three or more years in the case of large endoreheic basins like Lake Eyre.

Sabkhas—coastal plains situated just above high tide lines along desert coasts—are analogous to coastal plains in humid climates like the Chenier Flats in Louisiana. Although they share common elements such as sandbars, shorelines, stream systems and deflation hollows, they are mainly distinguished by carbonate sediments and minerals that remain after evaporation.

DUNES

SHAPES CREATED BY WIND

Overleaf: Moonrise over a Libyan sand sea in the Ténéré Desert.

Above: Cordons of dunes in the Libyan Desert form an elegant design.

The most common and universally accepted image of a desert is oceans of sand—formidable dune fields in motion. This generalization is somewhat valid but dune coverage of the planet's deserts actually is only around 10% of the deserts' total surface area. This does not minimize the importance of, and the drama inherent in, these wind-blown forms and the fascination that they have for geomorphologists and travelers alike.

Dunes consist of sand deposits made by wind (or eolian) action. Sand can be described as minute, irregularly shaped fragments of rock whose very constitution warehouses a huge number of geologic data, which permit the reconstruction of paleo-environmental, geographic, geomorphologic and terrestrial dynamics.

A grain of sand bears the signatures of the geologic events that produced the rock from which the grain eroded and the history, which, for the time being, has made it part of a desert and a dune before other events transform it by cementing it into sandstone. The

8

Characteristics	Study Method	Interpretation
mineralogical composition	polarizing microscope, x-ray diffraction	original rocks modified by alteration, transport
chemical composition	chemical analysis	original rocks modified by the environment, diagenesis
size	dryness, speed of sedimentation, microscopy	transport mechanisms that have acted on the original materials
shapes and roundness	microscopy	abrasion of the material during transport
surface texture	scanning microscope	alteration, environment and diagenesis
porosity, permeability, density	physical methods	environment and diagenesis

study of these particles is thus fundamentally important. An analysis includes the particles' crystalline structure, mineralogical composition (percentages of their component minerals), chemical composition, shape, age and the likely mechanical deformations that they have borne (see Table 8).

Sand is defined as any material from 0.0625–2 mm (0.0024–0.08 in) in size, independent of composition. Larger stone particles fall into the categories of grit and pebbles, while anything smaller is defined as silt; clays measure less than 0.0039 mm (0.000156 in) (see Table 9). Sand deposits assemble grains of different character, colors, sizes and shapes, and it is precisely these characteristics and their lack of homogeneity that allow us to track down the point of origin of their constituent rocks. There are essentially two kinds of sand: those whose components are not water-soluble (found in rain, rivers and oceans), and those that dissolve. The sands in the first group are made of silicates, which are the most widespread minerals on the planet. Quartz is clearly among the most abundant of these, followed by feldspars, iron, titanium oxide, hematite, magnetites and others.

Above: Fossilized dune formations of Antelope Canyon, U.S.A.

Below: Grains of quartz through an optical microscope; bottom: quartz crystals with various crystalline shapes.

9		
Granule Size (mm) [inches]		**Type of Granule**
>4	[>0.16]	pebbles—clumps
4–2	[0.16–0.08]	grit
2–1	[0.08–0.04]	coarse sand
0.5–0.25	[0.02–0.01]	medium sand
0.25–0.125	[0.01–0.005]	fine sand
0.125–0.0625	[0.005–0.0025]	very fine sand
0.0625–0.0039	[0.0025–0.000156]	silt
<0.0039	[<0.000156]	clay

Carbonate sands belong to the second group. Composed of calcite, aragonite and oölite, they are minute spherules composed of extremely thin concentric layers of calcium carbonate and, finally, by organic matter such as microscopic shell fragments. Among the most famous carbonate deposits are the Wahiba Sands of southern Oman. Deposits of intermediate materials may exist between the two main groups, as happens in New Mexico's famous White Sands dunes, for example, where the accumulations are made of gypsum granules.

The grains that sands are made

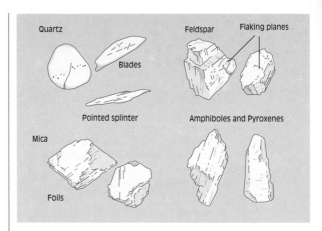

Quartz

Blades

Pointed splinter

Feldspar   Flaking planes

Amphiboles and Pyroxenes

Mica

Foils

Below: Grains of sand seen by electron microscope; bottom left: ripples in the white sand of the Australian desert; bottom right: differential erosion of sandstone and calcareous rocks in Egypt.

of are largely the result of the disintegration of pre-existing rocks, so they are effectively rocks in a different form, materials that have suffered both mechanical and chemical processes of alteration, splintering, disintegration, transport and re-deposition.

Shape is another useful trait that helps reconstruct the origin and history of sand deposits in general and of the single grains that compose them in particular. Generalizing, their shapes can be distinguished as jagged with deep angles and "cutting" facades, or rounded with blunt angles and smooth surfaces. From their shape, one can immediately deduce, in a general sense, how long the granules have undergone the mechanical effects of erosion and the distance they may have been transported. A grain of quartz must be transported thousands of miles before it starts to round off, or it may have trailed the same itinerary many times during the arc of geological time. A waterborne cube of quartz measuring 0.5 mm (0.02 in) per side would have to complete the equivalent of 50 trips around the world at the Equator before it could finally be rounded.

So it seems clear that sand made of quartzite, and thus particularly resistant and stable, can be involved in many cycles of erosion, deposition, interment, tectonic uprising and fresh erosion. Grains of quartz—as components of many well-rounded sands—can be testimony to numerous orogenetic and erosional cycles. The deposits they form are defined as "mature" sands, a term that implies the elimination of less-stable components and, consequently, an accumulation dominated by quartz.

Sediment shape is an attribute that can be determined by attentive observation, preferably with a magnifying lens. Shape can be rigorously defined by the ratio between the area of the transverse section of a grain and that of the smallest circle that will circumscribe it. Crumpled shapes and cylindrical, spherical, plate-, sheet- and cigar-like forms can be discerned.

Rounding is a characteristic that can be observed with great accuracy.

Above: The fragmentation of a sandstone stratum; continued weathering will turn it into sand.

Facing page and left: These diagrams illustrate the different forms (spherical, disk-like and cigar-shaped) that constituent crystals of sand grains of various mineralogical natures assume when they form sandy accumulations and dunes. The rounding of granules refers to their contour and edges. Shape and rounding are independent of one another.

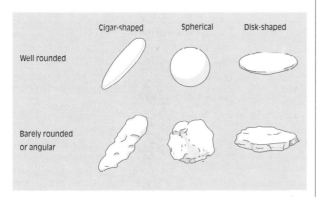

It is defined by the ratio between the average radius of the curve of its edges and that of the maximum circle inscribed. It expresses the wrinkling of a granule, and thus contemplates the presence of angles and edges (sketches on the following page). Rounding occurs and increases when grains collide with one another during transport and when they hit the ground, which tends to knock off any protruding corners.

Parameters have been established to define a grain's sphericality and others to define its rotundity. The quantitative assessment of the characteristics revealed is useful for creating classifications and, based upon them, defining the nature and origin of the deposit. The ability to observe sand with a scanning electron microscope has made it possible to capture extraordinary data and details that also help assemble—tessera by tessera—the geological history of these tiny universes in stone. Electron microscopes have revealed shapes and minute etchings that characterize such different environments

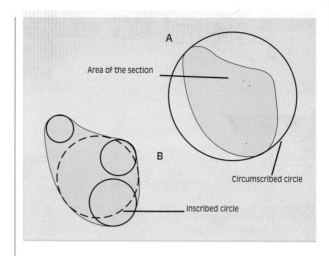

A

Area of the section

B

Circumscribed circle

Inscribed circle

of provenance as glacial, fluvial, marine or eolian, as well as the typical shapes and tread marks created when grains collide, which allow scientists to hypothesize the mechanisms that caused the displacement and transport of the sands themselves.

Wind is the agent most responsible for the movement of sand, clay and dust if we exclude transport by flowing water. Its capacity to capture, lift, carry and deposit solid particles is linked to its velocity. Wind flow is highly variable in both time and space. There are **constant winds** that blow with regularity at fixed intervals of time; **seasonal winds** that change direction often in the course of a day; and **inconstant winds** with irregular and sudden currents. The force and intensity of wind are then regulated and thwarted by the phenomena of attrition and turbulence, which manifest with irregularity on the earth's surface. Topographic, vegetation and other kinds of obstacles can change its direction. Once a current has succeeded in picking up a certain quantity of material, an action known as **deflation**, movement within the current occurs in **suspension** or by **saltation** (see page 52). In addition to wind intensity, the speed and quantity of material moved depend upon the weight, shape and size of the grains in transit. Lightweight material like dust, volcanic ash or fluvial-glacial silt can obviously reach great heights and can be transported great distances in suspension, even thousands of miles, in a regime of turbulent, intense winds.

Below: Electron microscope images of sand grains reveal the various shapes that each grain assumes according to its own evolutionary history.

Left: The formation of an immense sandstorm can be seen in this satellite image.

Below: Surface movement by saltation and suspension of sand grains in Australia.

Sand-sized material travels substantially shorter distances, bouncing along, never far from the surface in the process of saltation mentioned earlier. Some particles undergo multiple saltation hops. Others hop but once, their motion triggered by the nearby impact of a saltating grain, a process called **reptation**. The sand grains suffer damage during saltation, which can be observed under an electron microscope. If the surface material is too large or too heavy to saltate readily, it can experience displacement on the ground in a kind of chain reaction that mobilizes the entire surface of the terrain. This sort of movement of sand (or better, coarse sand, which assumes less importance) is called **rolling**. Particularly intense winds are capable of rolling particles as large as 5–8 cm (2–3 in), especially if there is small sand in motion striking the rolling particles. Wind thus operates selectively on sediments made of unconsolidated particles with various **granulometrics** including silt, fine sand, medium and coarse sand, and pebbles. The smallest fragments will naturally be the most easily entrained,

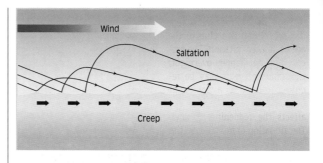

Right: Depending upon wind intensity, grains of sand can move by saltation, a series of long leaps, or by creep (reptation) across the surface.

Wind

Saltation

Creep

Below: The mechanism of grain movement by saltation. Trajectories over a uniformly sandy ground (A) and over a coarse ground (B) are shown.

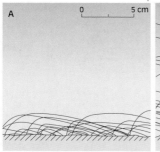

A   0   5 cm

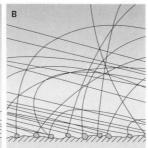

B

Below right: This sketch at bottom right illustrates the forces that act on a grain of sand deposited on a desert surface. The sketch at the top of the following page shows the creation of a lag of surface pebbles by eolian deflation.

Below: A grain of sand observed under an electron microscope. A scar caused by the impact of other grains is visible.

and a **sorted** surface will endure on the ground. It will be rich in heavy materials called the **deflation residual**, which protects underlying matter while the entire territory affected by the phenomenon is called a **deflation hollow**. These gravelly desert surfaces so distinct from sand accumulations are called **serir** and **reg**.

Color is another important and easily distinguished attribute of sand. It can vary from a brilliant white (whose primary and sometimes exclusive components can be rock salt, gypsum or limestone) through a wide gamut of yellows, yellow-orange, browns and bright red. Finally, some sands are quite dark due to a generous component of black grains of basalt or heavy minerals like magnetite.

Red is the most common color of sands that are closest to the surface. The sands themselves may contain a fraction of clay that can add to the dunes' red tint, and more rarely, fragments of red rocks such as flints will give off the reddish hue but, most commonly, iron oxides may completely envelop these grains.

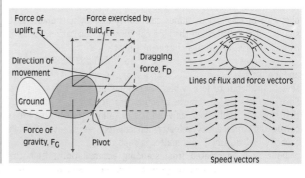

Force of uplift, F_L

Force exercised by fluid, F_F

Direction of movement

Dragging force, F_D

Ground

Force of gravity, F_G

Pivot

Lines of flux and force vectors

Speed vectors

The tendency for sand to redden with time is a clue to the age of an accumulation. The process does not happen overnight; thousands of years may be required to build up a significant iron oxide coating. Climate also affects the reddening of sand. The alteration of iron in hematite requires humidity, and the reddest dunes known are the coastal dunes in tropical climates like Namibia's. High temperatures probably also intensify the reddening. Mechanical stability augments the coloration of a dune; if an accumulation is growing and changing, considerable abrasion occurs and this scraping and laceration among the grains diminishes its color. Whatever the status of a dune, it is useful to remember that the chemical environment within it can often be determined by an analysis of its color.

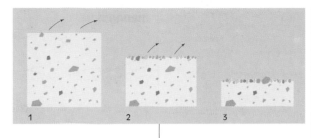

So sand forms distinct accumulations called dunes whose well-defined morphologies are typical of wind action.

HOW DUNES ARE FORMED

A dune is a landform made of sand—aggregates of clay and salt. Its height varies from 30 cm (12 in) to 400 m (1,300 ft), and it stretches from a meter to a kilometer (a yard to half a mile) wide. Its morphology is closely linked to constant wind action, while its potential for movement is bound to the topography of the substrate, including obstacles or vegetation. Dunes can slowly lose their characteristic shape and mobility when they are colonized by vegetation, and soil begins to form following a change of climate.

The size of a dune is defined by three elements: **height**, **width** and **length** (see drawing at bottom left on the opposite page). Dunes sometimes begin to form where a topographic surface is irregular. Small depressions or minute irregularities or even vegetation can first trap the sand. Then, as wind passes from a perfectly flat to a corrugated surface, its velocity immediately drops, its carrying capacity is diminished, and matter is deposited.

Dunes can be created from standing accumulations, from the sandy residue left by the passage of a mobile dune, and by the strong

Above left and right: The gritty surface of the Libyan Desert, and a desert surface mottled by nummulites and fossil foraminifers, in Egypt.

Below: Ripples produced by the wind on a dune in Australia.

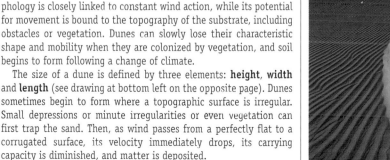

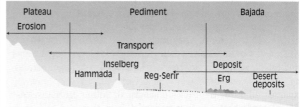

Above: A sea of dunes in the Algerian desert.

Above right: The schematic profile of a section of desert showing where erosion, transport and deposition occur; below: the elements of a dune; bottom: the mechanism by which a sand dune moves forward.

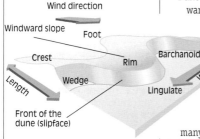

descendant currents that often occur in arid zones. Aimed at the ground, these currents sweep sand up and drop it elsewhere in the form of an accumulation.

Isolated and asymmetrical dunes are created when a wind bearing grains runs into a patch of sand in its way. The grains, which had been rebounding vigorously from the hard surface upwind of the patch, now are striking sand grains in the patch. The grains thus struck move, absorbing some of the energy that would have gone into the rebound of the impacting grain, which is slowed to the point that it stops within the patch. So many grains are now in motion that the momentum of the wind near the surface becomes too low to sustain grain motion, and the sand patch grows. The wind adds to its load by lifting particles from the patch. This higher volume and its added weight cause the load to be deposited and a mobile dune is formed.

Dune movement takes place when deflation occurs on the windward side, giving rise to a phenomenon of erosion. Sand is lifted and pushed upward along the slop by air flows until it reaches the crest, where the material falls, tumbling and accumulating on the leeward side, which usually has a 35° slope. The crest advances windward and the entire dune moves forward, keeping its shape and its asymmetry (illustrations at left and on the following page).

Enough factors combine to build a dune, and there are so many resulting types and shapes that the classification of dunes is difficult. The first and most important factor is wind—its intensity, regularity and distribution in the course of a day or a year.

Dunes that form in environments characterized by high wind energy will be larger than those that form where the wind is materially weaker. Still other factors are atmospheric stratification, topography of the substrate, the amount of humidity present, precipitation, and finally, the region's geologic configuration. Even the structure of "proto-dunes" helps determine their later development.

Several overriding factors make possible the creation of extended dune fields in general. They include the direction of prevailing winds, deflation zones along their itinerary where great quantities of sand can be picked up, constant and localized repetition of decreasing wind speeds, and finally, the presence of pre-existing sandy expanses. In 1973, the geomorphologist G. Wilson calculated that 85% of sands that are unconsolidated (and therefore subject to

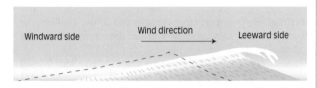

Left: A section of a transverse dune and its movement by deflation (the dotted section shows its location before the shift).

eolian transport and deposition) are found in accumulation zones otherwise known as **sand seas**. He defined sand seas as areas that are at least 30,000 sq km (11,580 sq mi) in magnitude, invariably covered by dunes and subject to constant winds, where the space between one accumulation and the next is no wider than the length of the natural wave of the dune itself.

Another attribute of dune fields is the presence of more than ten extended dunes in areas smaller than 30,000 sq km (11,580 sq mi). Forty-five percent of these areas are found in Asia, with 34% in Africa and 20% in Australia.

The illustration on the following page suggests a classification of dunes that takes multiple features into consideration.

Above left: Vegetation slows sand down; above right: an example of how vegetation has trapped enough sand to form a proto-dune.

PRINCIPAL DUNE TYPES

Obstacles that have been fastened in place by cementation and colonized by vegetation arrest and stabilize dune migration. The obstacles can be topographic such as rocks, hills and depressions, or vegetable, in which case the dunes are called "phytogenetic." These latter form around plants that must be at least 15 cm (6 in) high before they can seize and hold sand, sediments or clay. Called **nebkha**, a North African term, phytogenetic dunes are common and grow in size wherever winds are substantially weak. They grow until

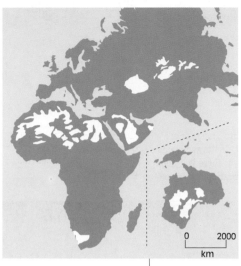

Above: The geographic distribution of active and stabilized sand seas.

Below: Various kinds of dunes and the prevailing winds responsible for their creation: A) transverse dune, B) barchan, C) longitudinal dune, D) parabolic dune.

0   2000
km

they reach a height where the wind's velocity is sufficiently strong to recapture a load of transported sand. The kind of plant that generates them determines their size. Bushes and shrubs build accumulations up to 5 m (16 ft) high with a radius of about 10 m (33 ft) (see drawing on the following page). There are remarkable nebkha 10 m (30 ft) high, more than a kilometer (0.6 mi) wide, and hundreds of years old.

Where vegetation is especially widespread, **parabolic** dunes called **garmada** take shape. The biggest ones are found in the Thar Desert of India and Pakistan; they are horseshoe or half-moon in form, with their concave side facing the wind. The arms of the parabola are stabilized by vegetation, while the thicker central part of the dune is blown downwind, creating a parabolic or hair-pin shape.

Stabilized dunes are sand accumulations that cannot be remobilized by the wind and are "blocked" by especially tall obstacles in the terrain or by accumulations that have been subject to cementation by mineralogical components of the dune itself. Vegetation is often the mechanism of stabilization.

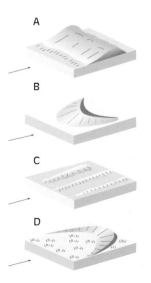

A

B

C

D

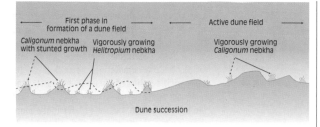

First phase in
formation of a dune field ← → ← Active dune field →

Caligonum nebkha   Vigorously growing   Vigorously growing
with stunted growth   *Helitropium* nebkha   *Caligonum* nebkha

Dune succession

Cementation can happen in response to a climatic change or, more frequently, during the accretion of the dune itself. If the sediments' clay content exceeds 15–20% of all matter present, there is a good chance that the dune will be anchored by disaggregation and cementation and will assume different shapes from those dunes composed of loose sand. These minuscule grains of clay may come from denuded surfaces, saline depressions or alluvial deposits, but largely derive from sabkhas and playas.

Above left and right: The creation of dunes anchored and stabilized by vegetation (also visible in the photograph).

Present on the surface of the dune, **clays** and salt-rich sediments dissolve when humidified and recrystallize as rigid crusts that ultimately stabilize the accumulations. The sands of eolian origin that suffer this kind of lithification are called eolianites.

Above: Various phases of anchored-dune formation due to native vegetation.

Far left: A solitary Tuareg in the first light of morning.

Mobile dunes are sand accumulations that can and typically move fairly rapidly and singularly or in groups across the desert surface. Among them, we distinguish **transverse** dunes, which lie perpendicular to the prevailing wind and have a rather low length-to-width ratio. Their windward slopes are gradual where sand advances by saltation and reptation. Their leeward slopes, called slipfaces, are steeper, where sand moves downslope due to gravity. They are asymmetrical in appearance with sinuous crests that document the existence of variable winds.

Transverse dunes often have on their flanks a network of small dunes formed by wind turbulence. Transverse dunes of various sizes occupy 40% of the sand seas in the world's various deserts, especially in China, the Sahel and Lake Chad. They may assume different morphologies, including **dome-shaped dunes**, where the transverse elements are incomplete. These are composed of fine sands, are practically free of slipfaces, and reach little more than 2 m (6.5 ft) in height.

Barchans, the type of dune most familiar to the public, take their name from the Turk word for "monument," but they are also known as **khet** in Qatar, **rabadh** in Saudi Arabia and **khord** in Mauritania.

They primarily develop in isolation on the stable surfaces of rocky deserts. In plan, they trace a crescent moon whose convex side faces the oncoming winds while its two extremities or wings trail to leeward. In section, their considerable lack of symmetry, which is typical of all transverse dunes, becomes apparent. These dunes can grow to marvelous heights of more than 100 m (330 ft) and widths close to 500 m (1,640 ft).

Barchans tend to lose sand from their wings as they move. Sand from the windward zone continually replaces the lost material so that the dune maintains its size and shape. The replenishment of sand is thus fundamental; if it were not available, the barchan would be destined to disappear. It has been calculated that a 3-m (10-ft) high dune in a Peruvian dune field employs 18 cubic meters (635 cubic feet) of sand every year in this continual cycle of loss and substitution. The same dune, then, is completely recycled every 64 years, having traveled a little over 1.7 km (1 mi). In places with the same eolian regime but more sand, barchans combine into huge landforms called transverse dunes that may span 3 km (1.8 mi) from wing to wing and more than 2 km (1.2 mi) in length.

Thirty to 60% of most sand seas—in the southern hemisphere, the Sahara and the Rub'al Khali—are occupied by **linear** dunes, (illustration at right). With a limited curvature and a symmetry that lend themselves to the formation of slipfaces on both sides, they may stretch more than 200 km (125 mi) long and, though rarely, 300 m (980 ft) high. The orientation of the linear dunes parallel to the wind prevents the dunes from migrating, even where they are not stabilized by vegetation. The most mobile element of a linear dune is its crest. Winds oblique to the dune can create sinuous undulations in the crest and cause their movement with time.

Linear dunes are quite regular in shape and a varied terminology exists to describe them. They are known as *sayf, seif, sif,* and *silk,*

Top: A barchan dune in the Libyan Desert; above: a dune supported against a rocky surface with distinct ripples.

Below: Barchan fields characteristic of the Libyan Desert.

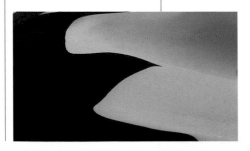

fluq, elb, zemoul, ghrud, uruq, tibba,
and *dune en vague*, while the valleys
between adjacent dunes are called
*goud, aftout, taieurt, feidj, straat,
ka'ar, omuramba* or corridor. Geo-
morphologists have attempted to
explain the processes that create
linear dunes by examining their principal origin, the repetition and maintenance
in space of the same form, their growth, parallelism and coexistence with other types of dunes.
Many hypotheses have been suggested, but a precise
codification may still be far off. Some theories propose topographic barriers and others the presence of complex wind regimes,
but these would only explain their evolution and not their origins.
Others think that the erosion of large pre-existing topographic structures causes linear dunes.

Colonization by plants can be caused by a large-scale climatic
change or by variations in precipitation or the eolian regime. These
changes take place over a wide span of time and are evident in the
Sahel dunes in western Africa, the Kalahari, a few areas of the Australian deserts and many other locations. Morphological studies have
shown that surfaces on the linear dunes' lowest slopes are more stable and protected and thus easier to colonize.

Large mobile dunes called **stellar** or **star** dunes mass up to 400 m
(1,310 ft) high at growth rates of 30 cm (12 in) and more, have
an almost unitary length-width ratio, and are characterized by slow
and limited movements. True mountains of sand, they are important elements in the desert landscape. They are also known as
oghrounds, ghords, barahis, medanos, or *demkhas,* and they are
certainly the tallest of mobile dunes. The most famous ones are
found in the Chinese deserts, the Lut in Iran, or those in Algeria's
Great Eastern Sand Sea, where they cover an area of 12,000 sq km
(4,630 sq mi). Star dunes assume a very regular distribution in space
and they occur in several variants—more elongated in one direction,
pyramidal with four arms, rounded off and compact with short layers, or dunes with three or more highly elongated arms. The origin
of these structures seems to be linked to multi-modal eolian regimes,
or winds that are effectively equal in intensity but blow from different and opposed directions and superimpose their load-bearing
capacities.

Mobile shapes in continual evolution, dunes animate the desert
and color it with a thousand shades. Sinuous crests with steep slopes
on one side and gentle on the other reflect sunlight artfully, creating shadows that give the landscape depth and width. Smaller structures that shape the slopes of the dunes themselves enormously
amplify these visual effects. Formed by secondary winds, the structures include complicated furrows and ripples, animal imprints

Top: The hypothesis of transverse-dune creation by turbulent winds.

Above: A satellite view of a field of linear dunes in the Algerian desert.

Above: Dune fields in the Algerian sand sea.

Below: A variety of ripples caused by wind blowing in several directions.

and traces left by the transit of bushes or tumbling pebbles. They form a boundless ensemble of ingredients that make sand deserts the most extraordinary places on the planet.

SAND SHEETS

Desert sands can accumulate in forms other than dunes. Broad deposits, or sand sheets, are essentially free of morphological elements; they are flat areas whose horizontal strata of sand are a few meters thick and practically undisturbed.

The Selima Sand Sheet is probably the largest. It touches the borders of Sudan, Libya and Egypt in an expanse of more than 100,000 sq km (38,610 sq mi). Others are found in Saudi Arabia, Namibia, the Ténéré Desert of Algeria and in some areas of the Australian deserts. Sand sheets often provide the base over which mobile dunes migrate. Sheet deposits seem to originate from at least four factors: fairly coarse sand with a median granular dimension of about 1.5 mm (0.06 in); vegetation that impedes the movement of sand and thus the formation of dunes; groundwater near the surface that emerges periodically to hinder dune formation; and finally, eolian erosion of pre-existing dune fields. Vegetation can also act as the nucleation point for the formation of dunes.

LOESS

This German term refers to an eolian deposit of silt from 0.01–0.05 mm (0.0004–0.002 in) and clay granules less than 0.002 mm (0.00008 in) in diameter. Abundant loess deposits are found in many parts of the world. In fact, up to 10% of the continental mass is covered to a considerable depth with this material, a fact that attests to cycles of dust storms repeated over thousands of years. Usually yellow-gray in color, loess can be lightly compacted at the moment of deposition but does not show the internal stratification that occurs in sand dunes. Loess is fine enough for winds to carry it in suspension, potentially crossing thousands of kilometers and passing from one continent to another. In areas with sufficient precipitation, loess hosts a distinctive erosion pattern called pinnate (feather-like) drainage.

The typical environments of deposition are those on the periphery and margins of deserts, steppes and prairies where herbaceous plants hold dust on the ground and runoff is infrequent. The world's most famous and extensive deposits are in the Huang-He basin in northern China. There, loess covers thousands of square kilometers to a thickness of 100 m (330 ft) and more. The silts come from Asia's central interior regions, and the deposition must have taken place throughout the Quaternary in order to build the loess deposits to the observed thickness. Other loess deposits are found in Argentina, central Europe (see map at left), North America (famous deposits in the Missouri-Mississippi Valley, Illinois, Nebraska and Kansas), and New Zealand.

Dark soils can develop on loess deposits. They are especially productive for cereals; the fertility of the southern Russian plains, Argentine Pampas, certain areas in China and the central United States are due to them.

Below: A satellite view of a loess deposit in the desert in China.

61

SHAPES CREATED BY WIND ABRASION

The wind has an enormous capacity to create and move great masses of dunes. Geological evidence such as huge deposits of dust and sand on ocean floors and on the Arctic and Antarctic ice sheets has shown how **winds** were more vigorous about 20,000 years ago, and their

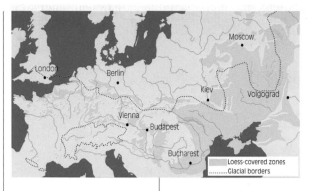

Above: The geographic distribution of Pleistocene loess deposits in Europe (pre-glacial period); the dotted lines indicate the positions reached by glaciers during their maximum expansion in the Pleistocene.

Below: Rocks of sandstone origin sculptured by abrasive winds in the Egyptian desert.

morphological action was more intense than it is today.

Desert winds are capable of carrying far more material than almost any other geomorphologic agent. Wind action in the Saharan Desert produces between 60 and 200 million tons of dust every year. By comparison, the River Niger—the only large river that drains the same area—transports 15 million tons every year.

Beyond its part in the formation of dune fields and dunes in general, and its intense erosive action by deflation (the entrainment of non-adhesive sediments), the wind is also responsible for the creation of unusual rock structures by **abrasion** (the consumption of cohesive, consolidated and resistant materials). **Abrasive action** occurs when the wind hurls grains of sand against solid rocks. When more or less spherical grains crash against a hard body, they produce fractures that are usually circular with a diameter equal to 10–30% that of the grain, depending upon the speed of impact. Repeated collisions increase the fragmentation and a crumbly surface is created. The kinetic energy dispersed on impact depends upon the diameter and speed of the grains and is therefore bound to the intensity of the wind. Sand is thus the principal abrasive agent but other sediments are significant, too. If the sand is angular, its erosive capability is even greater.

Studies conducted on the movement of grains of sand have shown that they do not achieve much altitude during transport. The sand's abrasive action is thus confined to between 10–40 cm (4–16 in) from the ground and diminishes as altitude increases.

Both sparse pebbles or rock masses can be objects of wind abrasion, causing typical and widely recognizable forms called ventifacts. On a small scale, pebbles and surface rocks are abraded into facets with distinct angles, some with minute sculptures, alveoli, pitting, holes, crests, grooves and slots. Grooves have closed terminals while slots are longer and open-ended.

When the grooves cross-cut rocks with different lithologies, they carve distinctive contours called **etchings**. Holes, indentations and protrusions result from the rock's heterogeneity, the presence of fractures and its lithologic diversity. The latter is important in furthering abrasion because the rock can only be abraded if it is resistant to weathering.

Wind-driven sand can also create more visible and larger morphologies, like some of the **mushroom rocks** produced when wind erodes a stem whose lithology may be different from, or more tender than, the top. In any case, meteorological and chemical weathering of the rock itself must also be considered in the formation of mushroom rocks.

YARDANGS

Yardangs are structures with dimensions spectacular enough to characterize landscapes on a regional scale. *Yardang* is a Turkish word used in geomorphology since 1903 to indicate a **promontory** that is deeply abraded by wind.

Yardangs show up as elongated hillocks parallel to one another and to the wind, owing their creation to its constant blowing. Substrate lithology is always important; it is usually a succession of

Top: Differential erosion of rock layers in the Libyan Desert; above: the softer portion is eroded and removed before the more resistant; differential erosion in Jordan.

Below: Quartz pebble lodged in less-resistant sandstone in Jordan.

Left: Preferential erosion of weaker sediments at the base of a stratigraphic sequence in Egypt.

sandy-clay sediments cemented in ancient lake deposits, but more resistant lithologies are also susceptible to this kind of abrasion.

Depending upon the nature of the rock, the force of the wind and the load of sand that grinds like an emery board, surface erosion ranges from less than 0.4 mm (0.02 in) to a maximum of a few centimeters per year. Yardangs are usually found in groups; their size varies from a meter to more than several kilometers, and their length-to-width ratio varies from 4 to 10. Larger yardangs range from 1–4 m (3–13 ft) high and up to hundreds of meters long; those found in Egypt, Chad, Mali and Algeria are typical.

They are strongly asymmetrical in profile and they usually appear in groups that cover wide areas, as mentioned earlier. The Kharga yardang field in Egypt extends across an area of 1,500 sq km (580 sq mi).

The most extraordinary yardangs occupy entire regions and feature exceptional dimensions. These structures extend as much as 650,000 sq km (251,000 sq mi), such as those found on the southwestern and southeastern flanks of the Tibesti Mountains in the central Sahara with the same northeast-to-southwest orientation as the Trade Winds.

Left: Two examples of deeply eroded surfaces, or yardangs, in the Eastern Egyptian Desert.

They reach more than 200 m (660 ft) high and kilometers in length, and stand more than 2 km (1.2 mi) apart. All this attests to a regime of high winds that are constant over time. Other yardangs have been mapped in Egypt and in Peru, and as high as 70 m (230 ft) in Iran.

DESERT SOILS

SURFACE MORPHOLOGIES

Soil is the outermost substrate of a stratigraphic sequence upon which life emerges. Different vegetable mixtures will develop upon it according to its chemical and mineralogical composition, and the vegetable mantle could not crop up in its absence. The granulometry of most desert **strata** or soils (otherwise known as surface **horizons**) is coarse due to the slow pace of weathering.

Desert soils are frequently characterized by horizons known as **duricrusts** because they typically abound in very hard minerals. Duricrusts are common in arid environments, but many of them took shape in ancient humid climates and are preserved in deserts because climatic conditions are so stable.

Many duricrusts owe their existence to processes of deposition, erosion and pedogenesis that occur in the course of complex climatic changes that last thousands if not millions of years. Sequences more than 200 m (650 ft) thick have been discovered and documented to contain as many as 45 overlaid duricrusts.

Duricrusts are an important element in desert morphology because they make up the primary contour in the flat deserts of central Australia, southern Africa and the central Sahara. Duricrusts are hard to wear down so they protect vast areas from erosion, and with the passing of time, these areas end up elevated with respect to the rest of a territory. For the same reason, duricrusts can also cause **inverted relief**. Where they once formed the base of a valley, erosion of the surrounding territories isolates the area protected by the mineralized crust, leading to the preservation of entire valleys that survive as elongated plateaus. Thus, ancient watersheds have become deep valleys, and the original promontories are inverted, a phenomenon common in Australia, Oman, Saudi Arabia, the central Sahara and China.

Overleaf: A pedestal rock in Egypt's White Desert.

Above: A surface horizon rich in salt; below: a stratigraphic sequence of saline horizons in Israel; below right: a saline horizon in the Negev Desert on the banks of the Dead Sea in Jordan.

68

DEEP SALT-LADEN HORIZONS

Salts such as chlorides, sulfates and nitrates are normally washed out of surface horizons; they can only persist on the surface in extremely arid desert conditions. Extensive deposits of sodium nitrate from a few centimeters up to a few meters thick are found in Peruvian and northern Chilean outcrops. These thicknesses imply the persistence of extremely arid conditions over long periods of time, and perhaps as much as 10–15 million years.

In slightly more humid conditions in southwestern Egypt, on the ground and in surface horizons, there are borates and carbonates of sodium, sodium sulfates and, in the deepest position, sodium chloride, calcium chloride, calcium nitrate and sodium nitrate. It is thought that these salt deposits may derive from the Neolithic era's somewhat more humid climate. The main source of salts in these cases is likely to be the salt content of dust from playa lakes and coastal zones. The source of Chilean nitrates is less apparent; they may arise from marine algae, guano, the bacterial decay of organic material or from volcanic rocks.

DEEP GYPSUM HORIZONS

Duricrusts rich in gypsum, or chalk, are called **gypcrete**. They may be 5 m (15 ft) thick and more but are not especially dense. They may form by pedogenesis starting with windborne gypsum dust from playas or, more rarely, from deposits rich in sulfates. Gypsum will also crystallize on the crown of the capillary fringe where a water table laden with calcium or sulfates is found in a surficial position. Some gypcretes feature large crystals, the famous **desert roses**. These jewels of nature are largely found in areas close to playas or sabkhas like the great Chotts in southern Tunisia and the beaches of the Persian Gulf. A meter-thick crust of this type took shape in less than 20 years in the Algerian Souf region after the **piezometric surface** of the water table was lowered when large quantities of water were extracted for irrigation. Some gypcretes will be created following the burial of playa evaporites.

Southern Tunisian gypcretes, which originate in the chalk dust produced by the desiccation of briny lakes, are about 8,000–9,500 years old. Those in the Dead Sea are believed to be more than 70,000 years old. Once exposed on the surface, they are too soluble to resist climatic changes and too weak to resist wind erosion, so they do not become permanent elements of the landscape like other duricrusts.

DEEP CARBONATE HORIZONS

Deep horizons rich in carbonates are prevalent in most of the soils on Earth, so much so that the quantity of limestone stored in them exceeds the amount found in terrestrial vegetation. These horizons

Below: Two images of a chalk-laden surface horizon in Algeria.

Above left: A horizon rich in marine fossils; right: a calcareous surface horizon in Egypt.

Below: A surface layer rich in silica in Saudi Arabia.

reach their maximum expression in arid or semiarid climates. Duricrusts made of calcium carbonates are called **calcretes**, **caliches**, or even **chalks**, while those composed of calcium or magnesium carbonates take the name **dolocrete**. They are generally a few meters thick, but there are known instances of 200-m (650-ft) thick calcretes. Many calcretes are pedological in origin, which means they are created in soils where the carbonates dissolve near the surface due to an elevated CO_2 content. Thereafter, following the infiltration of water, the carbonates migrate toward the bottom and precipitate to the base of the soil due to the diminishing concentration of CO_2. Calcretes of pedological origin form in more humid conditions than those found in today's deserts, and their presence in these latter is an indication of climatic variations.

There are also non-pedogenetic calcretes, which may be **phreatic** or **stream-borne**. The first are created by evaporation in the capillary fringe, and the second are caused by water flows rich in carbonates. The presence of these latter calcretes in deserts is once again indicative of significant climatic variations. Some calcretes in Oman and Australia were formed at the beginning of the Tertiary while stream calcretes incorporating paleolithic objects have been observed in northern Sudan, where no water now flows, even in the wadi, nor is there a water table near the surface.

DEEP SILICATE HORIZONS

Silicon oxide (SiO_2) is not very soluble in general but will rapidly enter into solution under the elevated pH conditions normally observed in arid soils. Such environments thus favor the precipitation of silicates by evaporation. This process occurs in other climates as well, as in the cold humidity of Wales, according to what degree the solubility of the silica is also controlled by the presence of other ions in solution. Duricrusts rich in silicates are called **silcretes** and are characterized by greater induration, a gray color, and a silicate component higher than 90%. Cement is made of microcrystalline quartz and opal.

There are three conditions under which silicates accumulate in desert soils: in shallow, recent profiles; in association with calcretes; and in sediments along the edges of playas. The first condition has been observed in U.S. soils rich in volcanic glasses where the silcretes present are made of uncemented nodules.

Above: Nodules of flint in a silicate-rich horizon.

Silcretes associated with calcretes are found in the Kalahari and southern Australia. They are the products of trade-offs that occur among different minerals under specific conditions. In sediments along the margins of playas, silica is released by the high pH and precipitates when solutions rich in silica come into contact with saline solutions beneath the playa. The largest and most widespread silcretes such as those in Australia, southern Africa and the Sahara are relics of earlier climates. In fact, silcretes in Australia began to form during the late Jurassic on stable low-rise landscapes and in hot humid environments.

IRON-RICH HORIZONS

Ferricretes are duricrusts containing up to 80% iron that can reach a thickness of 10 m (33 ft) and more. Relics of earlier climates, their structure may be massive, vesicular (blister-like), pisolitic (pea-shaped), nodular or layered. Various authors attribute the origin of ferricretes to pedogenesis and colluvial, alluvial, lacustrine or phreatic accumulation.

Ferricretes are generated in hot, humid environments with a strong seasonal component. Thus, when they are found in present-day deserts, ferricretes document climatic variations. Well-developed ferricretes are widespread along the edges of the Australian Desert, in the Sahel, and in the savanna region of western Africa; those farther north in the Sahara are thinner and less well-developed.

SOILS IN DEPRESSIONS

Soils other than duricrusts are found in deserts. Those common in depressions in arid regions include saline soils (solonchaks), hydromorphic soils and takyrs.

Saline soils or **solonchaks** cover about 7% of the Earth's land surface, according to the FAO classification, and almost all of them develop in arid zones. Some are very ancient; those in Mesopotamia date back about 6,000–7,000 years. Most of the salts come from the ocean or from marine and volcanic sedimentary rocks. There are also artificial solonchaks that form under two conditions and are far more extensive than the natural ones:

1. When irrigation water flows across soils, it feeds the capillary fringe of the water table near the surface, leading to the capture and concentration of salts on the surface. Solonchaks develop, initially taking shape in hollows and then spreading to surrounding areas. They are especially widespread in the Middle East (about 50% of Earth's saline soils) but are common wherever there is irrigation.
2. Where the water table rises because of logging or deforestation, desiccation and an increase in soil salinity can follow. This type of salinization is growing at an alarming rate and involves much of central Asia, Australia, the Great Plains of North America and southern Italy.

Solonchaks may display a white surface crust composed of crystallized salts a few centimeters thick. Sodium chloride is the most common, but sulfates, carbonates, nitrates and borates are also often present. Since organic material does not decompose under these saline conditions, however, the surface of many solonchaks is characteristically dark. Most plants cannot resist the high osmotic pressure on their roots in saline soils, so many solonchaks are free of vegetation. A few plants called halophytes adapt to saline conditions

Above: A surface horizon rich in salts in Israel.

and may contribute to increased salinity. Tamarisks, for example, raise salts from groundwater through their roots and release them on the surface. Solonchaks and other **hydromorphic soils** are generated when a depression is invaded at least seasonally by a rising water table and are characterized by the precipitation of various salts in the depths.

The Russian term **takyr** is used to indicate muddy soils found in the great depressions of central Asia. These have also been described

in the Middle East, however, and in North Africa. Takyrs are low in carbonates and are saline at depth; they exhibit a hard alkaline crust during the dry season while the crusts absorb water, swell, and become adhesive during rainy periods.

PAVEMENTS OF STONE

These structures have a limited thickness and are made of angular or rounded rock fragments. They are found on the surface or within a matrix of finer materials, composed of sand, silt or clay, and may occupy areas from a few square meters wide up to hundreds of hectares. They are usually found atop alluvial deposits and soils, in environments very different from one another including, for example, hot deserts. Radically diverse morphologies are on display in them that are as diverse as the names used to describe them around the world: *gibber plains* or *stony mantles* in Australia, *saï* in central Asia, *desert pavements* in North America, and *hammada* and *reg* in North Africa and the Middle East.

Stone pavements are common features of the desert environment. They stabilize the surface by furnishing protection to underlying desert soils. On a par with vegetation, they provide a barrier against the processes that attack the surface and decisively influence the mechanisms of surface infiltration. Once removed, these pavements are not quickly reconstructed, giving free reign to the accelerated erosion of the materials they once protected.

The stone mosaic is composed of multiple-sized rocks whose density and spacing is extremely variable, due both to differences in the original composition of the deposit and to the morphological phenomena that they have endured. As sloping increases, the local density of coarse particles grows while their spacing decreases. It also happens that the fine material occupying the spaces between large fragments contains a percentage of clay or salt. This leads to a surficial hardening that is thin but unyielding and provides for the deposition of the famous **patina** or **desert varnish**.

The variety of pavements is quite striking. As cited above, they may derive from materials of diverse origin, surface processes and various pedogenetics. The principal factors that concentrate rocks on the surface are bound to the deflation of fine matter, the removal of minute particles by surface rilling, or to complex operations that cause large particles to migrate toward the surface. Freeze-thaw,

Above left: A surface made of faceted rocks; below: a rock-strewn surface in the Libyan Desert.

73

Above: Wind-faceted rocks of various dimensions form a rocky pavement in the Libyan Desert.

74

hydration and desiccation cycles can induce the vertical migration of the soil's coarsest fragments and thus explain the concentration of rocks on the surface and in the depths of a sequence. The hydration/desiccation process operates as follows to generate pavements. Soil, particularly clay-rich soil, swells when **saturated**. Desiccation causes fractures to form; the smallest particles migrate into them, but the larger ones do not fit and thus find themselves carried to relatively higher positions. The mechanism primarily responsible for the formation of these gritty accumulations in hot deserts is surely deflation, which alters fine surface materials and leaves a residue of coarse particles. In the case of deflation, the final concentration of particles depends upon their distribution within the original sediment and the intensity of the deflation itself.

Soil processes and deflation are important in the creation and evolution of pavements, but they may not be the only causes. In Israel, for example, pavements are also created by the action of surface waters. Describing the factors that remove materials from the surface is complex. Other factors to consider are the often intense mechanical action of raindrops; the protective agency of surface crusts whose formation is accommodated by rain water; the impermeability of those crusts, which favors rilling; and increased permeability because of the fracturing of clayey minerals.

The relative importance of these processes and their interactions in the creation of pavements is far from clear. Even though it is likely that all these factors can operate independently in optimal circumstances, it would still be incorrect to state categorically that the operations described are the only ones that result in the formation of pavements. A hypothesis that does not rely on soil profile to infer the evolution of pavements was well illustrated on a series of 15 Holocene alluvial terraces near the Dead Sea. There, the degree of pavement growth (measured as a percentage of surfaces covered in stone) was more pronounced on older surfaces than on younger ones. Such changes correlate with about 14,000 years of meteorological alteration and contemporary differentiation of the soil profile. The entire group of alterations can furnish a relative and approximate means for dating surfaces whose age could not otherwise be determined.

It is uncommon to find pavements on surfaces that developed before the last glaciation or more recently. Whatever the processes and whatever their relative importance may be, it is clear that stone pavements, once stabilized, protect the underlying surface from erosion by water or wind and permit the soils' continued evolution.

GROUND PATTERNS

The phenomena of ground patterns—**circles**, **meshes**, **polygons**, **steps** and **stripes**—are well known in high latitudes and altitudes where they are generally explained by the action of **freezing** and **solifluction**. Similar corroboration is also found in other climatic environments and is fairly common in certain deserts. Though the results of freezing, whether current or historical, cannot be wholly ignored, earth patterning in the desert is normally thought to be the effect of two different and often correlated groups of processes that develop near the surface: hydration and desiccation, and the dissolution and recrystallization of salts.

The effects of the first pair of processes are typical of flat desert areas at low elevations and are conspicuous in fine sediments or in "inflated" or expansive soils, especially where these contain a high percentage of absorbent clays like montmorillonite. The effects of salt crystallization are evident in areas surrounding the edges of ephemeral bodies of water, along drainage canals, and where the capillary movement of water in the soil intercepts the surface.

The term *gilgai*, an Australian aboriginal word meaning "small water well," is applied to certain shallow surface undulations of clay-rich soils. Many of these appear to be linked to the differential expansion and contraction of soil coincident with periods of hydration and desiccation. *Gilgais* have been observed in many arid and semiarid areas, including Coober Pedy and New South Wales in Australia, the central Sahara, the Middle East, the black tropical soils of eastern and central Africa, South Dakota and elsewhere. They usually appear in soils whose clay content increases with depth and where, consequently, the inflation of subsurface soils is significantly greater than inflation on the surface.

Gilgai topography features a complex succession of swelling and depressions, canals and the

Below: Polygonal mud created by the contraction of clay in Jordan.

flat areas that fall among all these elements. Many theories about the origin of *gilgais* have been advanced and it is probable that they arise through several mechanisms. The type of *gilgai* created depends upon the depth of the level of maximum expansion and the variations of expansion among different levels. *Gilgais* can be circular or stair-stepped in shape, and in both cases the larger materials rest in border areas while finer ones lie in the center.

POLYGONAL SURFACES

These surfaces develop when a saturated, cohesive and fine-grained sediment dries out, passes through liquid, plastic and solid-fragile phases, and its volume is reduced to the limit of contraction. The microtopography of desiccated **sediment** will display systems of fissures and polygonal blocks. The fissures are slightly rectilinear or barely curved; and their length, depth, width, number and distribution vary enormously. Similar processes in arid and periglacial environments create these **fissures** inasmuch as both contractions are the result of an actual volumetric loss of available humidity. The number of fissures is inversely proportional to the granulometry. The spacing of the fissures can increase proportionally to the rate of desiccation and to the percentage of clay present in the material. The type of clay and the cohesive properties of the material will influence the magnitude of the contraction. The presence of rocks on a surface often determines the points of departure of the cracks. The chunks between the cracks may sport surfaces with concave, convex, flat or irregular profiles.

A concave surface is attributed to the fastest desiccation of the superficial level with respect to the material immediately beneath it. In its extreme form, the concave superficial level may create mud cracks.

Convex profiles may form when the fine-grained matrix contains a lot of salt.

Flat profiles will form when thick horizons of salt-poor materials dry slowly. There are two common partition schemes: orthogonal

Above: Polygonal mud cracks in the Jordanian Desert.

76

Left: Polygonal mud cracks caused by the desiccation of a muddy surface in Egypt; right: more polygonal mud cracks in various shapes and sizes.

Below left: A piece of dried surface mud; below right: camel prints in desiccated mud.

with right angles and tri-radial with 120° angles. Orthogonal systems are probably characteristic of materials that are not homogeneous or plastic in which the stimuli increase gradually, while non-orthogonal systems take shape within relatively non-plastic, highly heterogeneous materials that dry uniformly.

DESERT CHRONICLE

THE LANDSCAPE IN HISTORY

The desert's immobile and sometimes spectral image and its look of apparent lifelessness can suggest an immutable landscape that is beyond history, crystallized always and forever in time.

Its reality is something else. Many stylized figures that represent typical animals of the savanna are painted or, more often, scratched on rock walls. Millions of grinding stones and other prehistoric utensils are abandoned in many desert regions. They are the unequivocal records—from a not so remote past—of lush vegetation and abundant life in a place that enjoyed very different climatic conditions than those that exist today.

Today's arid regions were the theater of one of the greatest environmental upheavals to occur on the surface of Earth in the last ten thousand years. In fact, deserts that nearly disappeared a few millennia back have reconquered immense territory in a short time, and occasionally, in just a few hundred years.

The scale of this phenomenon is comparable only to what happened in temperate regions at the end of the last Ice Age when the glaciers began to withdraw. Unlike areas affected by glaciation, however, the desert regions were densely populated. It was precisely in those times that sheep-rearing first emerged in a part of Africa, while agriculture took hold and began to spread in Asia.

Overleaf: Prehistoric graffiti representing buffalo in Libya's Acacus Desert.

Above: Remnants of eroded rock structures in Libya and grinding stones in the Libyan Desert.

Right: A hard-to-read prehistoric etching.

Climate change and the ensuing aridity had dramatic effects on the human community, priming migrations, the extinction of cultures, and distinct kinds of cultural adaptation to the new environment.

The desert we see today, as illustrated in preceding chapters, is the result of a series of geomorphologic phenomena that primarily involve solar and wind energy, with water playing out a small secondary role. Working jointly, these primary factors govern the existence of the two known types of landscape: the sea of dunes (**erg** or **edeyen**) and rocky deserts (**hammada**). In them, typical features and characteristics can be identified that appear dissonant with the morphological processes in effect today, processes that began when the desert had more water.

Above, from left: Differential erosion on Jordanian sandstone (left and center); the Algerian sand sea; a stone pavement in the Libyan Desert.

At left: Mountainous promontories in Syria; earth pinnacles in Egypt; Uadi Rum in the Jordanian Desert.

MOUNTAIN MASSES

The central Saharan massifs (Tassili, Acacus, Messak, Hoggar) are characterized by certain geomorphologic features that date back to the late Tertiary and are the remnants of a tropical climate that prevailed at that time. These basic features are inselbergs and surrounding *pediments*, peneplains, and even the remains of ancient laterite soils and forests of rock pinnacles. Valleys and canyons also furrow these moun-

tain masses, which is evidence of ancient fluvial networks far larger than today's scarce water resources would yield.

The mountain landscape does not lend itself to preserving evidence of the humid climate in the Upper and Middle Holocene. Today's geomorphologic agents—erosion and the degradation of

From top, clockwise: Three rock formations caused by gravitational erosion; prehistoric caves and likely rock shelter in the Jordanian desert; entrance to a karstic cave in Egypt.

Below right: A schematic section of sand and rock desert areas showing sedimentary and paleosol deposits.

walls by gravitational action—tend to cancel more recent morphologic and sedimentary evidence, especially from the Quaternary, because these are generally made of unconsolidated deposits.

Karst cavities and spacious rock shelters came to be formed, however, and are the relics of Tertiary morphogenesis. They developed via the dissolution and hydrolysis of siliceous sandstone and originated at the same time as the so-called stratification joints and fractures caused by dissolution and rock falls (illustrated below). These places bear the most significance in a geologic study of the area since the caves have protected and preserved bits of deposit from erosion.

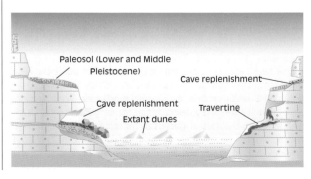

Paleosol (Lower and Middle Pleistocene)

Cave replenishment

Cave replenishment

Travertine

Extant dunes

THE PRESENCE OF HUMANS

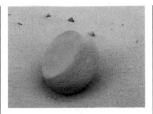

The last ten millennia of a region's climatic history can be reconstructed from an analysis of its terrains. The recovery of archeological data also lends weight to such studies since many deposits contain the traces of human passage in the form of stone structures, hearths and accretions of organic material that hold relics of the epoch's "artisans" including pottery, stone-age tools, grinding stones and other materials. Some of the most significant galleries of prehistoric art (especially Saharan) are found on the walls of these **caves** and **rock shelters** in the form of paintings and graffiti.

In Libya, there is a gorge or wadi called the Teshuinat that cuts across the Acacus Mountains. Travertine deposits in the form of rock slides and massive stalactite curtains that developed where two rock strata meet have been observed in some of the wadi's high valley caves.

Travertine is created by a sedimentary process typical of water-rich environments where water is concentrated in hydric cisterns within mountain masses following long periods of strong and constant rains. Conditions today are radically different; the Teshuinat is an extremely arid region with no more than 20 mm (0.75 in) of rainfall a year. The travertine appears to have formed between 10,000 and 14,000 years ago, indicating that the most intense period of rain in the Saharan region fell between the end of the Upper Pleistocene and the Holocene. But the more articulated stratigraphic successions are preserved in big caverns or caves that are left over from a vast Tertiary karst system. The famous Tin Hanakaten cave in the Algerian Tassili, for example, contains a 5-m (16-ft) thick stratigraphic succession and documents that humans frequented this place with regularity during the long interval of time from the Mesolithic to the Late Pastoral.

Ground sediments of eolic sands dated to the Upper Pleis-

tocene have been found in a stratigraphic sequence inside the huge Uan Afuda cavern and in the Uan Tabu rock shelter in Libya. These red-colored sands contain no organic substances. Their analysis shows that the area was a desert during the early Pleistocene, which coincides with the apex of glaciers in the middle latitudes, and did not much differ from today's environmental conditions. The sands contain chipped stone objects affiliated with the Musterian and Aterian cultures, which prove that humans frequented this desert

From top: An ancient grinding stone in Algeria; an ancient hearth in Libya; prehistoric art in Libya, and Paleolithic flint utensils recovered in Algeria.

Right: This diagram illustrates the chronological division of the Neolithic period, which came to be known as the "Pastoral" in the Sahara to indicate a society with a productive economy.

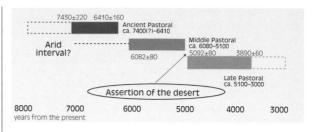

Below: The Acacus Desert in Libya, and an ancient settlement with a fireplace in the Libyan Desert.

area in Paleolithic times. The upper part of the stratigraphic sequence contains Holocene deposits that contrast with the older sediments in color, consistency and foreign matter.

Dark-colored deposits, instead, are rich in organic material and especially vegetal, ash and carbonaceous fragments linked to prehistoric hearths. Other than narrating the climatic status of epochs past, studying the caves reconstructs the events, customs and habits of ancient populations. Among the findings are windbreak shelters in Libya's **Uan Afuda** cavern, large hearths and posts from a habitation in Uan Tabu and drystone walls in the Tin Tohra cave. Large vegetal deposits dominated by the remains of herbaceous plants were neither burned nor associated with the fossilized feces of animals (coprolites). Such an accumulation of vegetable debris cannot be related to any natural process but was certainly carried into the cave by humans with the objective, in all likelihood, of laying up forage for animals. The large quantity of sheep and goat manure suggests that one of the most ancient attempts to domesticate these animals (among attempts that date back to the 10th millennium B.C.) happened in this cave. A type of goat, *Almotrago lervia*, still survives today but in a wild state in the Saharan massif.

Caves containing sediments that date back to the early Holocene or that contain relics of the Epipaleolithic and Mesolithic cultures are quite rare. Caverns containing remnants of pastoral civilizations are far more numerous and developed from the eighth to the third millennia B.C. (see illustration, preceding page). Their deposits are strongly influenced by the presence of humans. Containing hearths and other structures, they are rich in organic matter and vegetable remains that document the ongoing use of these shelters by a group of humans and the herds they bred. The same sequence is repeated quite systematically in stratigraphic successions. Black sands in the base layer contain profoundly decomposed organic matter, and few fragments of vegetable fiber are preserved. High in the succession, the number and state of preservation of the vegetal fragments increase, and the summit contains about 10 cm (4 in) of perfectly preserved goat droppings that contain whole vegetables, insects and other organic remains. Since these are generally lost over time, they provide a formidable source of paleoecological data.

Above: Stratigraphic sequence of the Uan Afuda cave:
1) collapsed boulders
2) unit 3—eolic sands
3) unit 2—colluvial sands and concretions in layered planes
4) unit 1—sand remains rich in decomposed organic matter
5) unit 1—rocks and ash lenses
6) unit 1—lenses of non-decomposed vegetal material
7) unit 0—eolic material and sand.

The **black sands** at the base of the succession require a humid environment to transform organic matter into humus; the circulation of solutions in the sediment also attests to the presence of patinas and chalky concretions. Only minimal or completely absent bacterial activity brought about by a constant environmental aridity can explain the perfect preservation of the manure found at the top of the sequence.

Identical sedimentary layers are common in the entire central Saharan caverns. **Palynologic stratigraphy**, or the analysis of pollens in the sediments studied in the Mathendush cavern, confirms this paleoclimatic tendency (see illustration on following page). In particular, the deposits (black sands) found at the base of the sequence contain pollens from the forested savanna and *Typha*, a plant that requires ample amounts of water. Moving toward the top, instead, and especially in the nearly surficial manure, the pollens of humid plants are progressively replaced by the vegetation of arid steppes. Today, of course, the environment is completely desertified and there is virtually no vegetation.

One typical feature of mountainous areas and the rock plateaus

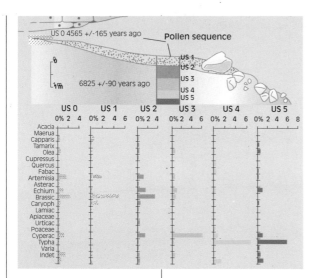

US 0 4565 +/-165 years ago

Pollen sequence

US 1
US 2
US 3
US 4
US 5

6825 +/-90 years ago

	US 0	US 1	US 2	US 3	US 4	US 5
	0% 2 4	0% 2 4 6	0% 2 4	0% 2 4 6	0% 2 4 6	0% 2 4 6 8
Acacia						
Maerua						
Capparis						
Tamarix						
Olea						
Cupressus						
Quercus						
Fabac						
Artemisia						
Asterac						
Echium						
Brassic						
Caryoph						
Lamiac						
Apiaceae						
Urticac						
Poaceae						
Cyperac						
Typha						
Varia						
Indet						

40 micron

86

called hammadas is the **black color** of surfaces created by desert varnish. As explained earlier (page 41), this is a very thin patina found in every desert in the world. It is a few microns thick and preferably develops on rocks that are most resistant to wind erosion like sandstone, lava, granite, and to a lesser extent, limestone. Its black color and sometimes-metallic appearance are due to the composition of the patina, whose primary components are manganese and iron hydroxides. Windborne dusts, and specifically the bacteria that concentrate manganese in the dust and in the rock, attach themselves to uneven rock surfaces to create desert varnish. These bacteria cannot live in overly humid conditions—which favor competitor organisms like lichen and moss—or in very arid conditions because they are suppressed by the alkalinity of dust and by the complete absence of water.

An exemplary locale for the development of a desert varnish is the Messak Plateau in the central Sahara, which appears entirely black since it is systematically covered by desert varnish despite its composition of light-colored sandstone.

The varnish in this area is ancient and is no longer being created. In fact, rocks that have fallen from the edges of the wadi in recent landslides and the cracks that furrow its walls and wind-abraded surfaces do not exhibit any patina.

Above: A schematic profile of the Mathendusc Cave; the diagram shows the percentage of pollens in the stratigraphic sequence.

Top: A section of desert patina or varnish rich in iron and manganese observed under an electron microscope; 1-2-3 are microstrata of patina deposited at successive times, 4 are grains of quartz.

Right: A desert surface with many rocks displaying desert varnish.

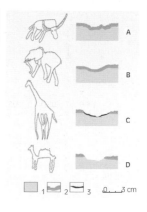

On this page: Prehistoric graffiti in the Acacus Desert in Libya depicting ancient fauna: ostriches, giraffes, crocodiles, elephants, and hippopotamuses (exemplified in the sketch).

Under microscopic examination, the structure of the desert varnish proves quite complex, consisting of three separate layers. The deepest of them is the surface of the rock altered by water. Above it lies the manganiferous patina, in turn covered by a very thin veil of quartz dust. Each of the three layers has a different environmental meaning.

The weathering of rock at the base of the patina required humid environmental conditions in which the walls could take in abundant precipitation. The manganiferous patina formed later, as aridity increased, and the last layer of dust was finally deposited under the auspices of the current desert conditions. By correlating the patina to carbon-dated funerary monuments, we know that it took shape between 4,500 and 5,000 years ago when the area was changing into a desert and the environment was transformed from a water-rich, forested savanna into an arid steppe that preceded the ultimate desertification.

The study of the desert varnish also carries significant consequences in the dating and

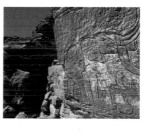

paleoenvironmental interpretation of prehistoric Saharan art, which has one of its major concentrations in the Messak. The oldest incisions depict large savanna animals including elephants, rhinoceroses, giraffes, ostriches, crocodiles and hippopotamuses, and their grooves are corroded and coated by the same dense black varnish found on natural rock surfaces. Etchings from the later **Ancient Pastoral Culture** also portray domestic animals but have a fresh, younger furrow; they are still covered with a black patina, which is a little lighter at times than that of the natural surfaces. Finally,

On this page: Prehistoric graffiti representing camels and hunt scenes in Jordan's Uadi Rum.

groups of etchings that belong to the **Recent Pastoral Culture**, which also portray domestic animals but with a more synoptic style, are free of the black varnish and veiled instead by a thin reddish alteration. More recent etchings from the protohistoric and historic ages, in which drawings of camels and warriors armed with shield and sword appear, along with writings in the Tifinagh alphabet, do not show any surface alteration.

Desert varnish signals a chronological and paleoenvironmental demarcation among the different groups of graffiti. The two oldest groups (Ancient Wild Fauna and Ancient Pastoral) belong to the beginning of the Holocene and were inscribed when the area still enjoyed a humid climate. In particular, the group of **Ancient Wild Fauna** etchings, with its corroded grooves, passed through a long period of humidity during which the walls of rock were subject to active weathering. The late pastoral graffiti lack a black patina and were etched after arid conditions had already been established and were no longer suited to patina formation. Numerous graffiti with a red patina indicate that, even after aridification occurred, the ancient transportation routes were not abandoned but continued to be frequented by groups of shepherds, even though climatic conditions were far more challenging.

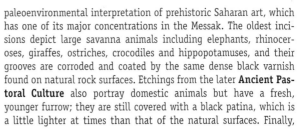

ERGS

Seas of dunes, or ergs, are the most characteristic feature of the desert landscape, even if they make up only a very reduced fraction of it. They are among the most ancient desert landforms as their formation dates to the beginnings of the desertification process toward the end of the Tertiary or the beginning of the Quaternary. In any case, the origin and age of the great dune systems are not known with certainty, but their diverse color and mineralogical composition leave room for hope that a detailed study may furnish more data. Ardito Desio, a famous centenar-

ian geologist (1897–2001) attributed the origin of the great erg in the Libyan Sahara to the deflation of the ancient Gulf of Sirt, which was abandoned by the ocean at the end of the Tertiary Period. Still, whenever iron-impregnated sand grains are observed, they prove to derive from the breakdown of laterite soils that advanced on the area's sandstone massifs (Tassili, Messak and Acacus) at the end of the Tertiary as an outcome of the tropical environment.

Below: Sand dunes in the Nigerian Ténéré. Bottom: A surface composed of fossil plant matter (probably roots) in the Eastern Egyptian Desert.

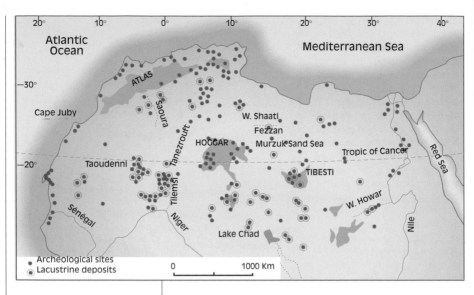

20°   10°   0°   10°   20°   30°   40°

Atlantic Ocean

Mediterranean Sea

−30°

ATLAS

Cape Juby

Saoura

Tanezrouft

W. Shaati

Fezzan

HOGGAR   Murzuk Sand Sea

Tropic of Cancer

Red Sea

−20°   Taoudenni

Tilemsi

TIBESTI

Sénégal

Niger

W. Howar

Nile

Lake Chad

• Archeological sites
◉ Lacustrine deposits

0   1000 Km

Above: Geographic position of prehistoric archeological sites related to the Upper Holocene period and of ancient lakes distributed in the Sahara Desert.

The dunes are, therefore, a stable feature and have cycled through more than one humid period. The interdune corridors between the great longitudinal dunes of the Murzuq ergs are terraced, and the terraces are composed of lacustrine deposits that contain artifacts (Acheulian) attributable to the Middle Pleistocene. Nonetheless, Holocene lacustrine deposits are the most widespread in the area. They have been observed in all the central Saharan ergs and edeyens (Murzuq, Ubari, Uan Kasa, etc.), but are generally reported in the literature of the entire Saharan region (Mali, Niger, Chad, Sudan, and Egypt). The same phenomenon is true of ergs in the Arabian, Indo-Pakistani and central Asian deserts.

The existence of ancient lakes or basins is thus well documented by the presence of characteristic deposits that endure in the interdune corridors or at the feet of the dunes themselves. They are fine sediments made of clay and silt whenever the lake had a tributary, and are normally imprinted by the roots of marshy plants and the

Below: Two large oases in the Libyan Desert.

obvious signs of stagnant water. Among the most characteristic sediments are the infinitesimal white limestone silts that contain a rich fauna of gastropods, indicating that the water in these lakes had a low salinity. The ancient lakeshores preserve an almost imperceptible step at the base of a dune's smooth slope. The environment is characterized by organic soils that evolved in a water-saturated environment or by peat deposits and swamps. Root imprints often perforate these deposits, which indicates the presence at their edges of lush vegetation. The bones of large savanna mammals—elephants, hippopotamuses and the remains of domestic cattle—are sometimes found in them.

Hundreds of **archeological sites** positioned around the central Saharan lakes document the strong attraction that these areas were for Epipaleolithic and Mesolithic hunters and for successive groups of shepherds that frequented their banks, maybe seasonally, but certainly very intensively. Some of these sites in the Uan Kasa and Murzuq ergs are 500 m (1,640 ft) long and contain traces of several hundred campfires as well as numerous storage pits that contain still more neatly arranged animal bones, unbroken earthenware containers, and well-stowed grindstones and mullers.

Even today, a few small lakes survive in the Ubari Erg in southwestern Libya, despite the extreme aridity. A dense concentration of grasses and palms growing at the edges of the lakeshore at the foot of the dunes furnishes a very concrete idea of the sand sea's landscape during the early and middle Holocene. The interdune lakes have neither tributaries nor outlets, however, and they appear where rainfall is elevated thanks to the capacity of sandy bodies to behave as actual water cisterns (see illustration on preceding page). This phenomenon is explained by the presence of innumerable minute

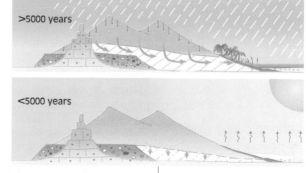

>5000 years

<5000 years

Top: A good example of a substrate rich in salts and chalk in Egypt.

Below: In the diagrams, the formation of an interdune lake and its later transformation by evaporation into a playa.

spaces among individual grains of sand. Like enormous sponges, the dunes thus capture and hold moisture in, creating suspended water tables that seep out at their edges through gravitation, and giving rise to hydromorphic soils and, where the topography is favorable, to lakeshores.

Lake water evaporates quickly in these arid regions. During this phase, certain alkalines (especially chalk, sodium carbonate and chlorides) are concentrated on top of the lacustrine deposits. Their precipitation creates the thick white crusts that characterize the roof of such deposits and render them easily identifiable on satellite photos.

Lacustrine deposits are easily dated with radiocarbon since they contain a large amount of organic matter. Thanks to this feature, the birth, extinction and oscillations of interdune lakes in the central Saharan are well understood. Lake levels rose progressively between

Top: A typical wadi in the Negev Desert in Israel; above: a large wadi in Syria.

8,000 and 8,500 years B.C., and after an abrupt interruption lasting a few centuries, went into another ascending phase in the seventh and eighth millennia B.C. The lakes dried out very rapidly around the fifth millennium B.C. in the central Saharan region, 500 years before desiccation invaded the lakes farther south in the sub-Saharan swath.

FLUVIAL PLAINS

Abandoned fluvial beds and the sedimentary deposits associated with them provide geomorphologic evidence that more explicitly documents periods of high rainfall. Though they are prevalent in every arid zone, the study of riverbeds poses special difficulties because they do not share the potential for dating that cave deposits and lacustrine sediments do. In some cases, though, and especially where they are associated with archeological remains, paleobeds can be dated and used to reconstruct the environmental events related to their activity and extinction (see chronological sequence in illustration below).

One example of a datable ancient watercourse is the Murgab Delta in Asia's Karakum Desert. The present oasis of Merv is found in the delta, which is now partially irrigated by an artificial canal (the Karakumski Canal) connected to the Aral Sea. Its northern section is desert-like and systematic studies and continued excavations since the 1950s have revealed that urban centers were present in full-blown desert from the Bronze and Iron Ages.

Only the proximal part of the Murgab's internal delta is affected by modern irrigation systems. Fine fluvial silt-and-clay deposits, sometimes interspersed with planes of muck, crop out of its distal section, which is sculpted by *yardangs* and lies below still-mobile dunes.

Geomorphologic observation has brought to light a tight network of ancient, inactive riverbeds characterized by meandering watercourses and cut lightly into the plains. Many archeological sites are distributed along traces of ancient man-made canals, among them

Below: A chronology of climate variations in the Messak Sattfet and Messak Mellet from the Pleistocene through modern times.

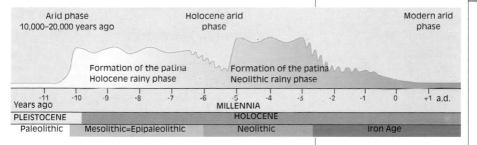

the fortified cities that rose around a kind of predominate palace, witness to a prosperous urban culture.

Radiocarbon dating attests to the existence of watercourses during the sixteenth century B.C. and again three centuries later during the development of Bronze Age cities, which is demonstrated by the direct connection between the sites and ancient fluvial beds. The urban center of Gonur, for example, is built on the margins of an ancient riverbed whose alluvium is clear and recognizable within the archeological stratigraphy of the Bronze Age. The presence of a man-made canal built to circumvent the city itself was also recognized. The city's abandonment coincides with the desiccation of the fluvial plain and the ensuing rapid advance of the desert from north to

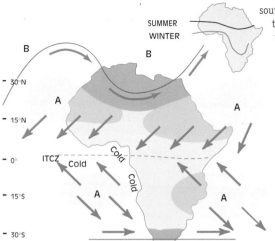

SUMMER
WINTER

B

B

30°N

A

A

15°N

0° ITCZ Cold

Cold

Cold

15°S A A

30°S

18,000 YEARS AGO

A model of atmospheric circulation related to a period that dates back 18,000 years. The dark area indicates a zone more humid than it is today, the shadowed zone instead shows zones drier than today, and the dotted line indicates the intertropical convergence zone.

south. It occurred gradually, first involving the Bronze Age sites and then small towns farther inland that were connected to the Iron Age. During the classical epoch, the desert began to lap up against the edges of the current Merv Oasis, which was fortified by an earthwork called the "Antioch Wall" that was built to defend the oasis against both the advancing desert and the nomadic communities that advanced with it.

DISAPPEARANCE OF THE DESERTS

Analysis and precision dating of the organic matter found in the desert and perfectly preserved by its hyperarid climate has permitted the reconstruction of much of the paleoclimatic evolution in many Saharan regions. A certain parallelism is observed between the middle latitudes and the temperate zones, the expansion of deserts and glaciations, and between the decline of desert areas and interglacial periods. The disappearance of the desert in the Saharan and sub-Saharan zones is in fact "contemporary" with the dissolution of the glaciers in Europe. After a long arid phase in the Sahara's central mountains, the beginning of an abrupt and noteworthy increase of moisture occurred, which is documented with certainty at 12,000 years, but may have started as long as 18,000 years ago.

The effects of this **climate change** were expressed in the birth of large and small lakes and widespread swamps. These transformations were felt first in the mountains and within a few millennia throughout the entire southern zone, which extends from Mali to the desert. The same phenomenon affected the lakes on the Ethiopian plateau and others in the subtropical fascia, including Lake Chad in particular, which started to expand at that moment and grew to a size that may have been comparable to the Caspian Sea.

The whole band of tropical deserts seems to have been characterized by the same climate. Humid conditions were documented throughout the Arab peninsula during these millennia—in Ramlat As Sabatayn in Yemen, in the Rub'al Kali and Waiba Sands (both became dotted with lakes), and in the Dhofar, where eolian sedimentation ceased and the mountains that border the coast were covered with thick vegetation. Still farther east, in the Thar Desert between Pakistan and India, the lakes that attest to humid conditions filled many interdune basins.

The extension of humid areas became a phenomenon that influenced the entire planet, and indirect but consistent traces of it exist, especially in arctic glaciers, which narrate a measurable increase in methane gas during this period. Fittingly, methane is the product of the immense biomass that develops in and around tropical lakes.

The first sign of **desiccation** appeared at the end of the ninth millennium with the rapid onset of a period of extreme aridity that lasted several centuries, as well documented in the central Sahara. Lakes in the sand sea dried out and many caverns emptied due to wind erosion. Such an episode must have involved a large part of the area now occupied by deserts since it was promptly registered in Antarctic glaciers by a decreased methane content and an increase in windborne dusts.

Another brief arid episode was documented in the Sahara at the end of the seventh millennium B.C., even though the seventh and eighth millennia had abundant precipitation throughout the area from the Atlantic coast, all of Africa north of the 20th parallel, Arabia up to the arid borders of India, the central Asian deserts, and perhaps even farther east (see illustration on preceding page).

Things changed abruptly in the following millennium. Around 5,000 years ago, the central Saharan mountain masses and the lakes in the surrounding sand sea dried out. Farther south, the conditions of aridity occurred in a slightly earlier time. The level of Ethiopian lakes dropped markedly about 4,500 years ago, while the great Taoudenni Lake in Mali dried out 3,800 years ago. Desertification of eastern regions is less understood; dates derived from sediments analysis indicates that aridification of the Thar Desert would date to the second millennium B.C., making it contemporaneous with desiccation of the Sahara.

Finally, paleoclimatologists the world over universally agree that the disappearance of deserts between the tenth and second millennia B.C. was primed by circulatory adjustments in the global atmosphere that occurred at the end of the glacial era due to the changing thermal situation caused by glacial melting.

During the Ice Age, seasonal fluctuations in the intertropical convergence zone (ITCZ) were greatly reduced since it was compressed at the lower latitudes by the presence of high pressures over Europe induced by the great glacial masses. With the disappearance of the glaciers and the fields of high pressure associated with them, the

Top: Well-preserved Paleolithic art found in Libya.

Above: Grinding stones and utensils in a Garamante village in Libya.

8,000–10,000 YEARS AGO

intertropical zone of convergence expanded up toward 20–25° north latitude, permitting the summer monsoon to carry rain from the Gulf of Guinea all the way to the central Sahara or even farther. Since the intertropical band where winds converge affected the whole planet, there was a general reinforcement of the humidity-charged southwest monsoons that penetrated as far as the deserts of Arabia and India.

The reasons for the environmental changes in Asia's continental deserts are more difficult to understand; they are even less understood and less well dated and directly influenced by the continental glaciers from adjacent mountain ranges and fluctuations of the giant interior lakes.

Above: A model of atmospheric circulation 8,000–10,000 years ago. The dark area indicates zones wetter than they are today; the shadowed area refers to areas drier then than now.

THE DESERT'S REVENGE

Starting in the third millennium B.C., the intertropical zone of convergence again stabilized, and thus returned to a "glacial" model of circulation. The Neoglacial period in Europe also began during this millennium. The hot and humid conditions of the Ipsithermic ended and the climate became more stable. The famous tribulations of the Similaun man, who was overrun and killed by an unexpected snowstorm and whose remains date back to 3,200–3,300 B.C., marked the beginning of this period in Italy.

Once the monsoon rains ceased, the desert rapidly reclaimed its lost territories and imposed radical changes on the subsistence strategies of the communities settled in these lands. Some of the pastoral communities were pushed from the central Sahara to its edges; some of them took refuge in the Nile basin and contributed to the development of the civilization of the pharaohs. Many others stayed in the deserts and found ways to adapt to the arid environment that persists today and is based on the oases and the network of caravans that link them. Some of these communities also evolve into governmental forms such as the Garamantes, the civilization of Napata, the southern Arabic kingdom, and so forth. In the Asiatic deserts, however, the imposition of aridity provoked the crisis and collapse of urban cultures and the affirmation of nomadic populations.

The path that the desert followed in its expansion is still far from being well known. From the fragmentary data currently available, desertification for many millennia spared those areas where ancient Holocene rains filled especially vast reservoirs that were not depleted by the monsoons' retraction to more southerly positions.

An example can be found in the central Saharan valley of the Uadi Tanezzuft, which survived at least into the first millennium B.C. thanks to the immense Tassili water aquifer found in the valley's drainage basin. Another contributor was the Uadi Darbat in Dhofar, which fed fresh water into the inlet where the city of Summurham sat on the Indian Ocean up until the second century A.D. thanks to the waters deposited in the karst network of Gebel Qara.

In an analogous way, the alimentation of the Murgab Delta and the processes tied to it lasted throughout the second millennium, after which the delta pulled back and the dunes of the Karakum progressively reclaimed its outer margin. Geo-

logic studies and tests have documented that the rains increased now and again, but a model of historic atmospheric circulation has yet to be reconstructed. A few groups of acacia in the central Sahara that date from the third century A.D. indicate a local resurgence of the aquifer; the Uadi Tanezzuft and Uadi el Ajal oases were still quite extensive in this period. This was the seat of the Garamantian kingdom. Aggravation of the desertification seems even more recent. Studies still under way indicate that dramatic soil erosion lowered the Uadi Tanezzuft alluvial plain by about 2 m (6.5 ft), generating a system of small dunes, an event that can be dated from 300 to 600 years ago. During those years, Ibn Battuda, the great Moroccan traveler of the 13th and 14th centuries, who united the coast of central Africa, was abandoned because he was crossing areas that were too arid. During the same period in Europe, the first Little Ice Age was under way, but this is a topic for later analyses.

Above: A large oasis in the Algerian Desert.

97

DESERT GUIDES

AFRICA AND EUROPE

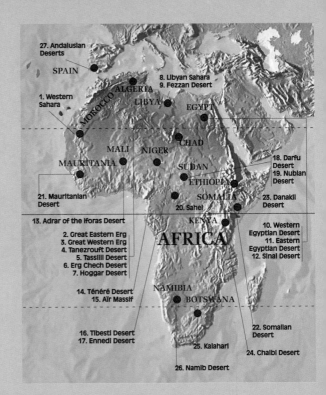

27. Andalusian Deserts
SPAIN
8. Libyan Sahara
9. Fezzan Desert
1. Western Sahara
ALGERIA
MOROCCO
LIBYA
EGYPT
MALI   NIGER   CHAD
MAURITANIA
SUDAN
18. Darfu Desert
19. Nublan Desert
ETHIOPIA
21. Mauritanian Desert
SOMALIA
23. Danakil Desert
20. Sahel
13. Adrar of the Iforas Desert
KENYA
10. Western Egyptian Desert
11. Eastern Egyptian Desert
12. Sinal Desert
2. Great Eastern Erg
3. Great Western Erg
4. Tanezrouft Desert
5. Tassilii Desert
6. Erg Chech Desert
7. Hoggar Desert
AFRICA
14. Ténéré Desert
15. Aïr Massif
NAMIBIA
BOTSWANA
16. Tibesti Desert
17. Ennedi Desert
22. Somalian Desert
25. Kalahari
24. Chalbi Desert
26. Namib Desert

Africa (along with the Arabian Peninsula) surely represents the desert continent *par excellence*, and everyone understands the Sahara as *the* desert.

Africa possesses two immense desert regions that are concentrated in the tropical zones north and south of the central equatorial zone. Established in the northern zone, the Sahara can be divided into several different regions, but it also manifests an unbroken evolutionary oneness in all of its magnitude from the Atlantic Ocean to the Red Sea.

The Sahara's distinctive morphology is inherited from its long geological evolution. The region was already flat during the Ordo-vician period between 450 and 500 million years ago when it was involved in a glaciation caused partially by the Sahara's geographic positioning at what then corresponded to the South Pole. Traces of this Ice Age are still visible, demonstrating that an environment generally maintains the footprints of consecutive climates, absent the obvious aftereffects of energy seen during a tectonic phase. From a tectonic point of view, the region remained stable for about 200 million years or nearly the whole span of time from the Triassic to the beginning of the Tertiary. It was under these conditions that an intense climatic change developed and produced the duricrust (see page 68), or distinctive soils that are visible in the present desert landscape. In the second phase of the Tertiary, tectonic activity in the form of orogenesis and volcanic activity in the Sahara gave rise to today's Hoggar and Tibesti massifs.

In the late Tertiary, tectonic movements between Africa and Europe led to the gradual reduction of the Tethys Sea and the formation of a great depression that corresponded precisely to the current boundaries of the Mediterranean. This occurred during the so-called "salinity crisis," which led to desiccation of the Mediterranean in the Messinian age (about 6 million years ago). The northern Sahara was involved in this desiccation and thick and broad sheets of evaporites (stratified deposits of rock salt and gypsum) formed on its surface. When the sea level dropped, the Nile carved a canyon 2,500 m (8,200 ft) deep, 1,300 km (800 mi) long, and 10–20 km (6–12 mi) wide. Later, the sea reoccupied the basin, reclaimed part of the northern Sahara, and penetrated as far as modern Sudan into the valley of the Nile.

Aridity came to dominate the Saharan environment starting about 3 million years ago. Sediment cores extracted from the bottom of the Atlantic offshore of the Sahara indicate that eolian processes were already in play at the end of the Cretaceous, or about 70 million years ago. This important variation in the environment was a consequence of the geologic and climatic changes that have affected the entire planet since the end of the Cretaceous. These include the opening of the Atlantic Ocean, uplift of the high Tibetan plateau, and the cooling of the oceans following the opening of the Drake Passage between Australia and South America.

Starting 2.6 million years ago with the beginning of the glaciation in the middle latitudes, the great tropical lakes within the Sahara began to dry up. Windborne sands invaded the Chad Basin and the vegetation changed decisively. During the last Ice Age, the sands of the Sahara also invaded the Atlantic's continental shelf. The lakes of the southern Sahara dried up between 16,000 and 23,000 years ago, and the active dunes pushed 450 km (280 mi) south toward the Sahel, blocking the course of the Senegal and Niger rivers. Toward the end of the last Ice Age about 12,500 years ago, environmental conditions changed yet again. The lakes in the Chad Basin in western Africa reappeared, and vast lakes and marshes also formed along the course of the White Nile and the Blue Nile. About 8,000–9,000 years ago, humidity increased markedly with a 65% increment in precipitation. Nonetheless, starting 5,000 years ago, conditions again began to deteriorate irregularly, tending toward the current situation.

The Namib and Kalahari are the two main deserts in southern Africa, and the Gran Karroo is a third, though semidesert, area.

The main attribute of the Namib coastal desert is a sand sea spread across 34,000 sq km (13,130 sq mi). The Namib Desert began to form in the Eocene and Oligocene (between 20 and 40 million years ago), when offshore ocean temperatures dropped. By the Upper Miocene (7–10 million years ago), the cold Benguela Current developed off southwest Africa and inaugurated hyperarid conditions in the Namib that continued into modern times despite a few short and not very intense episodes of increased humidity.

The Kalahari is an inland desert that has existed since the Cretaceous (65–135 million years ago). It experienced humid phases and the rise of pedogenesis (soil formation), which in turn produced the calcretes and silcretes still visible in the desert, and the formation of extended, but now fossil, drainage systems. Conditions during the Pleistocene were also more often arid than they are today (less than 150 mm (6 in) of annual rainfall). This is documented by the observation of fossil dune fields near the Victorian and Hwange Falls in Zimbabwe and in the Zambian, Angolan and Botswanan areas of the Kalahari, where annual average rainfall now exceeds 800 mm (30 in). There were also periods far more humid than today's, however, and great lakes such as the paleolake of the Makgadikgadi Depression in northern Botswana were formed. The maximum coverage of this lake was about 60,000 sq km (23,000 sq mi), about the same as modern Lake Victoria and bigger than the ancient Lake Bonneville in the United States.

Western Sahara

Nation:	Morocco
Expanse:	266,000 sq km (103,000 sq mi)
Average annual temperature:	10°C–27°C (50°F–80°F)
Rainfall:	<100 mm (<4 in/year)

Right: The "sand sea" and, on the facing page, layers of varicolored sandstone emerge from the desert plain near the Atlas Mountains.

The Western Sahara of Morocco is the stereotypical Sahara. Annual average temperatures range moderately from 10°–27°C (50° to 80°F), but thermal fluctuations are formidable at certain times of the year, with temperatures below freezing at night and over 50°C (122°F during the hottest daylight hours. The morphology of this part of the Sahara is a series of vast plateaus with tabular surfaces where vast rock zones spread up to the slopes of gentle promontories found in the south-central area.

The northern part of the desert occupies the Saguia el Hamra region and is bound on the north by the foothills of the imposing Atlas Mountains. Its eastern border is the Yetti Massif that rises partially in Algerian and partially in Mauritanian territory. It reaches the Atlantic Ocean on the west, and extends south into the region called the Gold Coast, which stretches as far as the Azèffal dunes and the Hammami Desert in Mauritania.

The rock formations that occupy the plateaus and characterize the landscape display morphological features that attest to intense and prolonged wind erosion, both abrasion and deflation. Blowing constantly in summer and winter, the Sirocco wind is the major agent for removing great quantities of sand and dust. These rock regions are rich in such mineral deposits as phosphates, iron, zinc, and to a lesser measure gold, which has begun to acquire an increased economic importance in the last several years. There are no surface waters in the territory yet narrow, deeply carved watercourses that indicate the dry beds of temporary rivers are evident. Referred to by the Arab term uadi (or wadi), these arroyos rapidly fill with water on the rare occasion of rain (usually less than 100 mm (4 in) a year), which is discharged in a single, short weather event. Numerous rivers flowing from the Atlas and other surrounding mountains are "taken captive" from the surface and run among impermeable layers of terrain protected from evaporation.

Occasionally, the hypogean river meets and intersects the topographical surface and emerges to create small oases whose waters are used to irrigate gardens that produce excellent products. Beyond

tains gradually disappear toward the southeast without abrupt passages or striking morphological and climate changes. Rocky, furrowed by a few wadi, and featuring high aridity and daily thermal fluctuations, the Dràa comprises a third of the Western Sahara in area. It shares a southwestern border with the Sahara, while bordering the Yetti Massif, Tindouf Desert and the salty Algerian Tindouf bog on the east. The

Above: Moonrise on the summit of a barchan.

Right: A sea of barchanoid dunes occupies the central part of the desert, completely free of vegetation.

these limited fertile areas, the Western Sahara is not completely lacking in vegetation. *Wariona saharae, Perraldeira coronopfiloia,* and *Trichodesma calcarata* are all found in the rocky areas near the uplands. In the transition zones that gradually lead to the desert and steppe areas, there are various spiny shrubs; Ziziphus lotus is one of the most widespread. Among the animals found there are antelope, gazelles, jackals, hyenas and desert foxes.

The Dràa Hammada extends along the Western Sahara, where the Atlas Moun-

surfaces on this desert's characteristic plateaus are empty of sand and other unconsolidated matter. Its rocks are deeply wind-abraded; they rest directly on the substrate and bedrock and always in a surficial position due to erosion. Precipitation is extremely rare and concentrated into brief episodes, which results in the creation of wadi. The most important of these is the Uadi Dràa that descends from the Atlas mountain chain and flows for about 1,100 km (680 mi) marking the border with Algeria. The river creates several oases along its run, among which the most important are the Ouarzazate and the Zagora.

Great Eastern Erg

Nation:	Algeria
Expanse:	192,000 sq km (74,131 sq mi)
Average annual temperature:	15°C–38°C (59°F–100°F)
Rainfall:	<70 mm (<2.75 in)/year

Clockwise: The crest of a stellar or star dune; ripples shaped by winds blowing on the windward surface of the dunes; dune fields; mobile dunes in the central desert: the "sand sea."

The Great Eastern Erg constitutes the central part of the Sahara. More than 70% of its surface, double that of the Great Western Erg, is occupied by a sea of sand. The desert lies east of the western Erg, from which a rock plateau nearly 100 km (60 mi) long separates it. It is bordered on the northeast by the Ksour Mountains in Tunisian territory, by the Tademait Plateau on the south, and by the Tinrhert Hammada on the southeast.

Sand dunes are the dominant morphology in this part of the Sahara and the climate is truly extreme. The wind regime that affects the entire region is especially varied and complex; the trade winds blow northeasterly throughout the summer, while cyclonic winds blow southwest to northwest at other times. Locally, violent currents of ascending hot air and unstable winds that concentrate in the area of the dunes and the surrounding mountainous uplands are also created. The winds accumulate the sand into cordons of dunes.

The scholar of desert morphology, Ian Wilson, confirmed that the Great Eastern Erg's sea of dunes evolves from the deflation of alluvial materials that surround the entire region. According to Wilson, the amount of out-

side sand reaching the Erg each year is a phenomenal 6 million tons. Calculating the average dune at around 117 m (383 ft) high and 26 m (85 ft) wide and counting the annual influx, it can be reckoned that the Algerian Desert has been accumulating sand at

that rate for at least 1,350,000 years. This enormous mass is in continuous movement in a constant direction. The height of the dunes increases toward the edges of the Erg, and the height disparity may exceed 150 m (492 ft) toward its innermost regions border-

ing the Tinrhert Hammada. In the outer zones, instead, spectacular star dunes take shape due to turbulent winds. They are smaller and closer together than the great crescent dunes. They vary from 0.7 km (0.4 mi) to 1.7 km (1 mi) in diameter, while the distance between two consecutive crests is not constant, falling between 0.8 km (0.5 mi) and 6.7 km (4 mi) on the northern border and less in the south, between 1.5–3.1 km (1–2 mi).

Great Western Erg

Nation:	Algeria
Expanse:	31,200 square miles
Average annual temperature:	15°C–38°C (59°F–100°F)
Rainfall:	<70 mm (<2.75 in)/year

Above: An imposing crescent dune over 250 m (820 ft) high.

Facing page: Shadows play on the crests of barchanoid dunes.

The Great Western Erg is Algeria's second "sand sea" after the Great Eastern Erg. Its territory extends into the country's northwestern region, shares its south-southwest border with the great Tademait Plateau, the northwest with the imposing Atlas mountain chain, and pushes west toward the Erg er Raoui. The extreme scarcity of precipitation at less than 70 mm (2.75 in) a year makes the climate difficult enough that animal and plant survival is sorely limited. The absence of soil, instability of the substrate, and the immense store of sand permit such plants as *Aristida pungens, Aristida plumosa,*

Tenutana and the *Lotus Jolyi* to grow on the dunes. Small reptiles, insects and formidable scorpions are always hiding among the high dunes. Humidity is very low and solar heating is intense, which often results in daytime temperatures that exceed 50°C (122°F).

The shape and organization of the dunes within the Erg depend on the local and regional prevailing winds. With stable, constant and unidirectional winds, sand distributes itself into large crescent dunes with smooth flanks and thin, apparently sharp crests. Under stable winds, dunes position themselves in parallel ridges sepa-

3

rated by wide hollows. Depending upon the force of the wind, these can travel as much as 30 m (100 ft) in one year. Where the wind is unstable, the dunes assume irregular shapes and tend to pile up on each other in a disorderly fashion. Violent winds that sweep across the entire Erg without meeting obstacles to slow them down are typical of this desert. Once they reach a certain velocity, the winds can capture and hold great quantities of material in suspension. Sandstorms are then kindled that may set up a front of almost 500 km (310 mi)—an impenetrable cloud capable of suffocating anything, including sound and light—and move at about 50 km/h (30 mi/h). The coarsest sand cannot get airborne above 1 or 1.5 m (3 or 5 ft), but is almost a natural emery board wielding considerable erosional power. Beyond this band, there is a layer of finer sand mixed with dust and minute grains of clay. Sand will often overrun the borders of the Sahara Desert during these events and cross the Mediterranean Sea into southern Europe and sometimes as far as Great Britain. A violent sandstorm in 1947 deposited a huge quantity of silts and clays on the snow-covered slopes of the Swiss Alps, coloring them an unusual pink.

Extreme conditions in the Algerian Desert have not permitted the establishment of human settlements. There is not one village in its interior and no roadways cross it. There are numerous small oases along the edges of the Great Western Erg, including the spectacular Taghit Oasis surrounded by an immense ring of red-orange dunes with date palms bordering the pond. Timimoun Oasis rises among yellow-ochre and red dunes where a village constructed in red sandstone has grown; the ancient Ksar fortress stands in the shadow of a palm grove; trunks of petrified wood may be admired in its environs.

Tanezrouft Desert

Nation:	Algeria
Expanse:	150,000 sq km (57,900 sq mi)
Average annual temperature:	14°C–37°C (57°F –99°F)
Rainfall:	<60 mm (<2.33 in)/year

Above: An ancient grinding stone from the Neolithic Age rises from the sand.

Below right: Flint arrowheads, also from the Neolithic Age.

Facing page: A desert section with no sand where livestock breeding is possible but precarious.

The Tanezrouft Desert lies in the south central region of Algeria. Some of it was occupied by a large lake basin that belonged to a complicated and extensive aquatic network that grooved the Sahara about 10,000 years ago. The process of climate change that began during that epoch led rapidly to a phase of progressive warming initiated in the Holocene. Rich in varicolored sandstone, the desert extends east from the Hoggar Mountains shaped by ancient metamorphic rocks. Defined on the northwest by the Erg Chech in Algerian territory and on the southwest by the El Knàchich Desert in Mali, this great plateau is dominated by the *reg*, or rock desert, composed of stones whose size and lithology vary. The territory's morphology is characterized by the mushroom rocks often associated with basins and depressions where wind action is greater on segments of less-resistant rock. Other unusual forms are the walls, mantles and pavements of rather dark-colored, eroded sandstone. Numerous canyons between 250–500 m (820 and 1,640 ft) deep make up its topography. They are more abundant in the fringe and surrounding zones. Sand dunes are extremely rare, while certain rock structures of sedimentary Paleozoic origin are common and characteristic of this desert. They appear in peculiar elliptical forms strongly eroded by the wind and fixed in tight contact with one another. Their com-

posite surface is so thoroughly sandblasted and burnished by the scouring action of the wind that the rocks come to be defined as desert "pavement." Colored Mesozoic limestone is visible in various areas; they are typical of the Sahara and cover the oldest sedimentary sequences. Below these layers lies the Precambrian African shield, strongly affected by folding on a regional scale. A large quantity of "fossil" water was found some hundreds of meters below the ancient bedrock, which may prove to be an exceptional hydrous resource in the future. The sand bears a characteristic red color due to the accumulation of ferrous oxides in the rocks from which it was formed. The climatic change that enveloped this zone also radically changed its principal erosive agent from water to wind, which in turn led to the progressive reduction of soils and, subsequently, of vegetation. The great scarcity of precipitation at no more than 60 mm (2.3 in) a year and a daily thermal fluctuation of 50°C (122°F) permit the growth of only exceptionally resistant plants like acacia and the jojoba tree. The whole of this Algerian desert is known as the "land of terror" for its extreme water deficiency, yet caravans have traveled north-south tracks from Gao and Timbuktu to Reggane for centuries. These ancient routes still exist in the desert, but big trucks cruise new roadways that are not bound to the locations of water wells. The local fauna includes jackals, hyenas, antelope, rabbits, gazelles and reptiles.

Tassili N'Ajjer Desert

Nation:	Algeria
Expanse:	80,000 sq km (31,000 sq mi)
Average annual temperature:	20.3°C (68.5°F)
Rainfall:	<30 mm (<1.2 in)/year

Right: A pile of stones in the Tassili Massif.

Far right: A Tuareg walks among this desert's stone pinnacles.

zones. It includes the mountainous region of Adrar, where Monte Isser rises to 2,254 m (7,395 ft). The entire mountainous region is made of Ordovician and Devonian sandstone; it surrounds the older crystalline Hoggar Massif, which dates to the Precambrian. The Tassili Plateau owes its structural morphol-

The Tassili is a mountainous region located in the central Sahara in southern Algeria, almost on the border with Libya, Niger and Morocco. It borders the Irharharene region on the north, the Hoggar massif to the northwest, Hammada Mangueni on the southeast, the Tafassesset Plains and Tarazit Massif in Niger, and finally adjoins the Murzuq Desert in Libya on the east. The Tassili is a plateau between 1,500 m (4,920 ft) high at its northernmost section and up to 1,800 m (5,900 ft) in the central and southern

ogy, including its steep valleys, to the repeated succession of humid and dry periods during earlier epochs. At the end of the Pleistocene, this region featured many large lakes fed by mountain streams. The entire area became extremely arid except for a few more humid episodes in the last 10,000 years. In some

areas, its morphology was shaped by fluvial action that left narrow ravines and numerous wadis. Elsewhere, wind erosion and the arid climate created vast expanses coated in rock formations known as the "stone forest." The plateau's southern section heaved up high enough to form an escarpment 600 m

(1,970 ft) high that divides the Tassili from surrounding ergs. Recent volcanic deposits covered the sandstone base and gave rise to the Adrar Mountains.

Much of the Tassili is considered hyperarid; any water is found in small quantities in the deepest rocky ravines. But there are also semiarid

Below: One of many ancient Neolithic graffiti in the Massif illustrates the desert's ancient fauna.

Facing page, from the top: Rock masses produced by erosion; prehistoric painting of a hunt scene; forest of stone pinnacles typical of the Tassili landscape.

areas, which are characterized by a fauna of lynx, antelope, eagles, reptiles, spiders and insects, and flora typical of the Mediterranean. Annual rainfall is only 30 mm (1.2 in) and the yearly temperature revolves around 20°C (68°F).

The extraordinary cypress, *Cupressus dupreziana*, is one of the three most important local plant species. It is found in the higher reaches of the plateau, such as the Valley of the Cypresses, where there are a hundred ancient trees, some probably millenarian and almost living fossils. There are also wild olives and myrtle, especially within the wadis.

In 1982, the entire region was included in the list of areas protected by UNESCO for its international geological and biological importance. Twenty-eight of its species are threatened with extinction. The Tassilis is also important

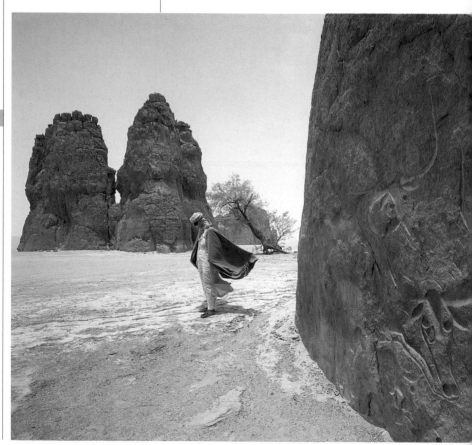

from an anthropological perspective. Many interior districts are known for the ancient paintings found on their cave walls. The most important prehistoric remains include drawings etched into rocks; the remains of Neolithic artifacts, sculptures and pottery that have been radiocarbon-dated to about 6,000 or 7,000 years ago; and objects from the middle Paleolithic. The area counted less than 3,000 inhabitants in 1986

and is even more sparsely inhabited today. Only a few nomadic and settled Tuareg pursue agriculture in the vicinity of a few wadis. The majority of the population is concentrated in oases like Djanet, which is the most important.

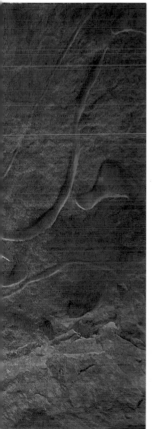

Erg Chech Desert

Nation:	Algeria
Expanse:	120,000 sq km (46,330 sq mi)
Average annual temperature:	25°C (77°F)
Rainfall:	25 mm (1 in)/year

Right: Mass of mobile dunes as high as 200 m (660 ft).

Facing page: A majestic compound dune.

The Erg Chech lies east of the Yetti Eglab Mountains in the middle of the Sahara. The Tademait Plateau separates it from the Great Western Erg and the Great Eastern Erg by a patch of sand and the low hills of the Ougarta Range. It borders the Tanezrouft Plateau to the south and El Eglab on the northwest, which in effect separates it from the Erg Iguidi.

Parallel geomorphologic structures and complex linear dunes make up the Erg Chech. These are disposed along a southwest bearing and are asymmetrical in shape. The side east of the crest is larger than the western side, and they are shorter in the western segment. The rock substrate is visible and crops out in the dark areas between successive dunes. The sandy linear structures are typically over 1,000 m (3,280 ft)

long and separated from one another by an average distance of a little over 5.5 km (3.4 mi), an unusually large separation for dunes of this type. Scholars attribute the

breadth of the interdune corridors to a singular scarcity of available sand. The negative balance of sand in the Erg reflects the fact that much dust and sand is exported from the region by the wind, much of which is deposited in the form of dunes or sand masses in the Sahel in Mali, or trapped against the slopes of the Atlas Mountains. The deposits originating from this Erg are eas-

ily recognizable and identified by the peculiar red coloration of their sands.

An aerial view of the region clarifies how the materials of Erg Chech and other small sandy deposits like the Erg Iguidi and the Erg Raoui invade the slopes and ravines of the Yetti Eglab massif. This is especially true of the Erg Raoui in particular, which is composed of a thin layer of sand deposited in the form of linear dunes and is positioned northeast of the main Erg between two lines of dark hills that extend northwest and southeast. The Yetti Eglab Massif, which seems to pull sand in its direction, is a hilly structure formed by a few granite rocks concentrated in the west and northeast zone. More than 600 million years old, the highly eroded, granite bedrock of the Precambrian shield emerges just north of these formations.

The fossil-rich dark rocks found in the southeast zone are younger and date to the Cambrian and Ordovician periods. Finally, the center of the massif is composed of Paleozoic rock.

Climate and environmental conditions in the Erg Chech reflect those of other ergs in this and adjacent regions, so rainfall is scarce, temperatures are high, and day-to-night thermal fluctuations are sharp. Flora and fauna are scarce and highly adapted to survive in this very hostile environment.

Hoggar Desert

Nation:	Algeria
Expanse:	80,000 sq km (31,000 sq mi)
Average annual temperature:	5°C–30°C (41°F–86°F)
Rainfall:	25 mm (1 in)/year

Above and on the facing page: Silhouettes of the basalt and granite that constitute the lithology of the Hoggar Massif.

The imposing Hoggar Mountains spread through the middle of the Sahara in Algerian territory atop the Tropic of Cancer. Arid and rocky, it is practically uninhabited, even by nomads who rarely venture through its narrow passes. Vegetation is also scarce; the few hardy species such as *Potamogeton hoogarensis, Silene hoogarensis, Senecio* and *Lupinus tassilicus* survive thanks to short and sporadic rains. In some especially protected and deep passes, water evaporates slowly enough to allow small pools of stagnant water to form where herbaceous plants proliferate. Among its peaks, Mount Tahat rises 2,918 m (9,574 ft) and its snows never melt in areas protected from the sun. The Hoggar Mountains meet the Tanezrouft plateau in Morocco on the northwest and are surrounded on the south and northwest by the Tassili Mountains. Daytime temperatures do not soar because of a high average altitude, but the daily thermal fluctuation is high, and the thermometer often drops below freezing at night.

Igneous in origin, the Hoggars alternate between intrusive rock formations, such as the granite at the base of the stratigraphic sequence, and extrusive rocks like the basalt that flowed abundantly from numerous volcanic craters at higher elevations. This basalt covers the partially eroded, ancient granite to a thickness

of about 150 m (500 ft), forming a rough surface characterized by sharp and erratically shaped volcanic slag. The Mount Ilama area is of particular scenic interest. A volcanic edifice with very steep slopes, it is superimposed upon pre-existing lava formations in a sequence of perfectly hexagonal columnar basalt called "the Mosque of Tamergidan." Given its uniqueness, this locale has been favored as a place of congregation and worship since the most ancient (Neolithic) times.

A sequence of volcanic rocks called phonolites is

found near the Mount Ilama area. The phonolites fracture along defined planes and give rise to tower- and spire-shaped structures or to narrow prisms that lean against each other or stand isolated like tubes of rock. More than 300 of these rock monoliths stretch across a surface a little larger than 775 sq km (300 sq mi) in the Atakor area.

Another area with an interesting morphology and landscape is the Assekrem Plateau. Among its pale sands one distinguishes a series of rock reefs, spires, overhangs, compact volcanic deposits and bedrock of intrusive ori-

Below: Rock structures of intrusive and extrusive origin eroded by weathering agents.

Facing page: Obvious grooves, the result of millennia of erosion, score the Hoggar's rock surfaces.

gin colored black by desert varnish. The famous Tamanrasset Oasis lies at the foot of the 1,400-m (4,600-ft) tall mountains to the southwest. Known by its nickname, *Tam-Tam*, it is a place of rest along the interminable Trans-Saharan Highway. It is also the refuge of numerous Tuareg nomads, the tribe of veiled and mysterious blue men who are one of the most enigmatic populations in Africa. Tuareg women enjoy a freedom that is unique within the continent's cultural panorama. TamTam has grown to 45,000 inhabitants, many of them descendants of slaves

as well as a good number of northern Algerians. The first house of Father de Focauld, a missionary and legendary desert figure who was the first among twentieth-century Europeans to approach the Tuareg culture seeking to study its language and customs, is at TamTam.

The fascinating Hoggar Mountains harbor many prehistoric grottos and caves where extraordinarily beautiful paintings have been found that date to between 2,000 and 8,000 years B.C. The pictures represent herds of cattle, rhinoceros, elephants, giraffes and jackals, animals that still live in Africa but far away from these places.

Spiral geometric symbols have been found among the graffiti that evoke rites of ancient civilizations of hunters and herdsmen.

Libyan Sahara

Nation:	Libya
Expanse:	1,200,000 sq km (463,320 sq mi)
Average annual temperature:	10°C–35°C (50°F–95°F)
Rainfall:	<100 mm (<4 in)/ year

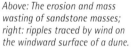

Above: The erosion and mass wasting of sandstone masses; right: ripples traced by wind on the windward surface of a dune.

Facing page: The interdune Mandara Lake, found deep within the Libyan sand sea.

The northeastern portion of the huge Sahara is called the Libyan Sahara. It extends from eastern Libya and coasts along the southeastern territory of Egypt until it reaches the northwestern extremities of Sudan at the Hammada al-Hamra. It is a plateau composed of reddish marl, sandstone and limestone, which give rise to its nickname, "Red Hammada." Many gorges are carved into the local rock whose black-colored surface is crusted with iron and manganese salts in an obvious desert varnish. The origin of the gorges is probably the erosive phenomena linked to cycles of freezing and thawing. The Libyan Sahara's typically monotonous landscape stretches beyond the Hammada al-Hamra plateau. It is fashioned of rock plateaus free of vegetation and extended sandy plains that are arid and inhospitable to most living beings.

The limestone tableland slowly drops toward a group of oases, at 29° latitude North, and extends south into the Serir Kalansho and the Girabub Erg, which are both extremely arid and hard to navigate. Continuing toward the extreme southern segment of the Libyan Sahara, one finds another characteristic rock plateau, which serves as a floor for the dazzling Cufra and Gialo oases.

Cufra, also called Al-Kufra, includes a group of oases situated in one of the typically elliptical depressions found in southeastern Libya, this one measuring 48 km (30 mi) long and 19 km (12 mi) wide. The oases lie on two sides of an ancient caravan route, and up until 1895, when they became the headquarters of the Sanusi religious community, they provided a stronghold against marauders. The most important oases towns are Al-Jawf, At-Tulaylib, At-Tallab and At-Taj. At-Taj rises on a rock plateau whose aquifer at a depth of 7–8 m (23–26 ft) permits the cultivation of cereals, olives, grapes and palm groves, as well as the breeding of sheep and goats. Gialo is the main city in the Gialo-Awjila group of oases. This system, which also includes the Jhurra Oasis, is composed of a valley about 30 km (19 mi) long that widens out here and there to form small shallow salt lakes. Extensive palm groves at the Gialo Oasis provide dates that are considered the best in Libya.

The oases disappear toward the Egyptian border and give way to an immense expanse of sand interrupted locally by rock massifs made of crystalline schist and granite that reach their maximum height around the 1,934 m (6,344 ft) high Jebel al-Uweinat. The climate is especially arid in this part of the Sahara, and more than 200 consecutive days of dryness have been registered in many areas. Fortunately, a hot wind called the Ghibli blows, which, though hot, seems to mitigate the arid and inhospitable desert climate. The vegetative cover is scarce under these conditions, even though some species of plants typical of semiarid regions can grow in the valleys carved by ephemeral streams. The fauna includes rodents, desert foxes, scorpions, insects and snakes.

123

Fezzan Desert

Nation:	Libya
Expanse:	551,170 sq km (212,806 sq mi)
Average annual temperature:	10°C–30°C (50°F–86°F)
Rainfall:	<10 mm (<0.4 in)/year

The Fezzan or Fazzan is a vast tract of the Sahara situated in southwest Libya. An administrative district with its capital at Sebha between 1951 and 1963, the territory borders Algeria on the west and Niger and Chad on the south. It merges with the eastern Libyan Desert on the east and Tripolitania on the north.

The territory looks like an immense hollow made of alternating sandstone and limestone horizontal layers that shape the great elongated plateaus or the stony substrate hidden beneath the dune fields. One of the most important physiographic elements of this desert is the Tadrat Acacus Massif whose north-south length exceeds 100 km (62 mi). The stratified rock that emerges from the sand at the base of these uplands dates back 400 million years to the Silurian and Devonian periods and is rich in fossils. To the west there is a steep vertical escarpment that drops 600 m (1,970 ft), while on the opposite, eastern face, the massif softly and gradually submerges under the sandy expanses with a mildly inclined plain. This fascinating region is rich in the artifacts of herders who lived during the Neolithic period in natural shelters created by fluvial erosion. Their ancient graffiti depict bovines (oxen, sheep, goats, antelopes), pastoral scenes and savanna animals. The Ghat Oasis lies within this zone typified by natural arches. It sits on the Algerian border and can be identified from afar by an ancient Turk fortress built above a spur of rock from which the palm groves and cliffs of Acacus may be admired. Another physiographic aspect typical of the Fezzan is the Erg of Uan Kasa. It features a system of well-developed linear dunes in the west, while uniting in the east with a thin blanket of sand. There are finely layered sediments of fluvial origin at the base of the dunes. They embody the remains of an environment different from today's that dates to the middle Pleistocene and contains Acheulian artifacts from 80,000 years ago. The Murzuq sand sea is the largest expanse of dunes in the whole Sahara, with northwest-southwest trending longitudinal dune crests 10–15 m (33–50 ft) long and 1.3–1.5 km (0.8–1 mi) wide. Interdune corridors about 2 km (1.2 mi) wide separate the crests. The discovery

Facing page: Natural arches, tafoni and spectacular erosional structures characterize the landscape of the Acacus.

Below: Caves, ravines and natural arches recur throughout the northern Messak.

Facing page: A stone tower more than 40 m (130 ft) tall in the Libyan Acacus is the product of weathering and mass wasting.

of crocodile, hippopotamus, elephant and gazelle bones, and the existence of fluvial deposits, confirm the presence of temporary bodies of water in the sand sea and their successive desiccation as a consequence of the climate change.

The Titersine Erg and Messak Settafet Plateau are also important geomorphic features. This latter is a rock plateau whose origin can be traced to an ancient erosional surface of the Tertiary. A network of serpentine wadi and canals spans the plateau. Recent studies have shown the presence of tropical palaeosols that submerge under the dunes of the Murzuq Erg in the east.

The Fezzan is characterized by a climate of hot summers and relatively cool

winters. Rainfall is scarce and does not exceed 10 mm (0.4 in) a year; some precipitation is seen in its northern district. The dry, hot Ghibli blows in spring and influences the climate of the entire region, driving temperatures over 50°C (122°F). The wind may last a few hours or many days and is capable of carrying huge quantities of sand. Nonethe-

less, the climate permits the existence of animals such as gazelles, fennec (small desert foxes), snakes and scorpions concentrated in the small but luxuriant oases distributed through the territory. The most important is the Sebha Oasis, where date palms, figs, and oleander grow.

Eastern Egyptian Desert

Nation:	Egypt
Expanse:	221,940 sq km (85,690 sq mi)
Average annual temperature:	9°C–41°C (48°F–106°F)
Rainfall:	<50 mm (<2 in)/year

Above: The calcareous (limestone) substrate crops out at the base of a sand dune.

Right: Unusual limestone structures built by differential erosive action.

Facing page: An example of bare white calcareous substrate emerging at the base of the dunes.

The Eastern or Arabian Desert, known in Arabic as the *as-Sahara' al-Gharbiyah,* occupies a quarter of Egypt and spreads between the Gulf of Suez, the Suez Canal and the Red Sea. Moving north to south, three geologically and geomorphologically distinct areas may be distinguished. The northern region is a plateau of limestone origin composed of soft

of 2,500 m (8,200 ft). At their base, the outlets of some of the major wadi form a few deep inlets, which support small settlements of semi-nomadic peoples.

A sandstone plateau extending south from Qina shapes the second geographic region. Its profile is highly irregular because of the numerous steep escarpments that characterize its morphology. The third region encompasses the coastal plain and a complex of hills beside the Red Sea. These are actually mountain chains that extend north to south and occupy the area between the Suez

hills that stretch from the coastal plain at the edge of the Mediterranean as far as Qina, a city on the banks of the Nile River. Near Qina, the plateau becomes a typical palisade deeply cut through by wadi that make it hard to cross. The cliffs reach a height

and the Sudanese border.

The mountains comprise separate, discontinuous systems that reach impressive heights such as the 2,187 m (7,175 ft) of Mount Shaiyb al-Banat. The geomorphologic formations found in this area are made of igneous and

metamorphic rocks and have a complex geological structure. Granite outcroppings near Aswan stretch into the Nile Valley to form the First Cataract, which is the first series of rapids visible going up the river.

Most of the valleys that typify this western region of the desert drain toward the Nile. They are usually dry, even though their distinctive structures were produced by the erosive agency of water and present irrefutable confirmation of climate change throughout the entire region. During the most humid season, the valleys turn into temporary watercourses.

Sandstorms called *khamsin* are typical of these regions

of eastern Egypt. They happen often between March and June because of the northeasterly movement of the low-pressure system found in Sudan. A sudden increase in temperature from 8°C to 11°C (46°F to 52°F) and a drop in relative humidity accompany the very dense, dusty and high-intensity wind currents of a khamsin. Rains are scarce in the desert's eastern regions and only permit the growth of a few plants able to withstand these conditions. They include tamarisk, acacia, spiny shrubs and aromatic herbs. Vultures, eagles and scorpions represent wild fauna. The nomadic communities abundantly distributed throughout the area live by

raising herds of sheep, camels and goats in this less prohibitive climate than that registered in the Western Desert.

Western Egyptian Desert

Nation:	Egypt
Expanse:	422,845 sq km (169,138 sq mi)
Average annual temperature:	10°C–30°C (50°F–86°F)
Rainfall:	<50 mm (<2 in)/year

130

Above: The Dakhla Oasis, famous for its sulfur springs.

Facing page: A bare surface free of soil, vegetation and sand in the southern part of the Western Desert.

The Western Desert, or *as-Sahara' al-Gharbiyah* in Arabic, occupies two-thirds of the surface of Egyptian lands. The Nile River borders it on the east, with Sudan on the south and the great Libyan Desert on the west. The boundaries between these two desert areas are not well defined. Their morphology is similar and does not offer relevant distinctions. The desert looks like a completely arid limestone plateau that lacks any kind of vegetation. Numerous and rather isolated small hills of sand document the wind's powerful capacity for erosion. *Khamsin* are also typical of the region.

The region offers spectac-ular geologic formations and unique and distinctive landforms such as the White Desert and numerous depressions or topographic basins whose surface lies below sea level. These latter extend for hundreds of kilometers and possess their own hydrographic network, which guarantees the development of oases within them. The most important depression is called Qattara; it occupies a surface of about 18,100 sq km (7,000 sq mi) at 133 m (436 ft) below sea level. Characteristic elements are an extraordinary salt lake and a few fields of natural methane gas that are important for the region's economy.

The Western Desert is characterized by the development of flourishing oases, among which El-Kharga is probably the most important. It is one of the biggest Egyptian oases and is situated in a depression 34 m (112 ft) below sea level. It hosts an archeological museum where discoveries from various ancient sites in the area may be experienced. The temple of Ibis may be visited north of the city of El-Kharga, the regional capital.

The temple is dedicated to the god Amon and surrounded by a lush grove of palms. Going north, there are the cemeteries of El-Bagawat and the ruins of an ancient monastery. The temple of Nadura rises on a hilltop to the east; from this summit, one can admire the spectacular landscape surrounding the city of Ibis. Dakhla is one of the most important oases for international tourism, in part because it hosts the Ethnographic Museum and remains of the ancient citadel. A spring of sulfurous waters is one of the main attractions of the oasis. It is ideal for exploiting the medicinal benefits of sulfur baths as one soaks in the 46°C (116°F) water that issues here.

The biggest Egyptian oasis is Siwa, which spreads through a depression 12 m (40 ft) below sea level in the center-west desert. Like a mirage in the expanse of sand, its beauty seems to have no equal in the vast Saharan panorama. Siwa is surrounded by green olive plantations irrigated by more than 300 springs that have created a multitude of salt lakes, which provide refuge for many species of birds. Ruins of the El Shali fortress stand out in the city center, and the remains of the citadel and Temple of Amon can be seen to the east. The Gebel el Mawta, or "Hill of the Dead," where numerous tombs of the Roman era have been preserved, is eloquent. The indisputable main attraction of the oasis is a lush tropical garden called Fatna.

The last important oasis is El-Fayum southwest of Cairo. It is a splendid area about 80 km (50 mi) wide east to west and 56 km (35 mi) north to south. Qaroun Lake is also special, and at 215 sq km (83 sq mi), is one of the largest salt lakes in Egypt. Many nomad and Bedouin encampments and small villages of fishermen are found on its banks. Together with Farafra, these oases belong to a new province called the New Valley.

Sinai Desert

Nation:	Egypt
Expanse:	61,000 sq km (23,550 sq mi)
Average annual temperature:	10°C–30°C (50°F–86°F)
Rainfall:	<100 mm (<4 in)/year

The name of the Sinai probably derives from the moon god *Sin* who was worshiped in this region in antiquity or from the Hebrew word *Shen*, the name of a typical bush in the region. It is a triangular peninsula that marks the point of union between the African and Asiatic continents. It stretches west between the Suez Gulf and the Suez Canal, which separates it from the Eastern Desert, and is confined on the east by the Gulf of Aqaba and the Negev Desert in Israeli territory. To the north, it reaches as far as the Mediterranean and slides into the Red Sea on the south.

Mountains rising from the desert plateau of the Sinai can top 2,500 m (8,200 ft); Mount Catherine, depicted here, is 2,642 m (8,668 ft) high.

12

The Sinai belongs to the arid climate belt that crosses northern Africa and southern Asia. The effects of this climate are evident in the high index of degradation of the ground surface, the presence of expansive dunes, areas distinguished by thick saline crusts, and by a multitude of dry wadi. Two main geomorphologic areas may be discerned. The first is a mountainous complex in the south-ern peninsula. Its igneous rocks are heavily incised by deep wadi oriented to drain toward either the Gulf of Suez or that of Aqaba. The eastern front of the mountains in this part of the desert rises almost abruptly and may stretch to significant heights, like Mount Catherine whose pinnacle tops 2,642 m (8,668 ft); Umm Shawmar at 2,585 m (8,481 ft); Atha-Thabt at 2,437 m (7,995 ft); and Mount Sinai with its 2,285 m (7,497 ft) above sea level. This is part of a protected area that also includes the Monastery of Saint Catherine, built by the Byzantine Emperor Justinian the First, and today is the object of pilgrimage. Due to its mountainous nature, clouds stand still in this territory as though bridled by the pinnacles, at least for the better part of the year. There are even snowstorms here

Above: Sandstone eroded and sanded down by the wind offers a spectacular play of colors.

Below right: A dry wadi near the Red Sea.

Facing page: Two of the most important peaks in the desert, Mount Sinai and Umm Shawmar.

during the winter while summer rains are influenced by monsoon winds blowing from the east. From the western flank a narrow coastal plain spreads out and borders the Gulf of Suez.

The second morphological region of the Sinai occupies two-thirds of the entire Sinai Peninsula. Here, a wide plateau dissipates into the Mediterranean, sloping gently from a height of more than 900 m

(2,950 ft). The plain of the Al-Arish Uadi, tall cathedrals of sand, and dune fields moving across coastal plains typify this platform. Intense winter rains characterize the "Mediterranean" climate in this region, while the summer is dry and torrid. A seasonal dry wind called the *khamsin* blows in the spring and fall and there are violent but only occasional rainstorms.

The southern Sinai Peninsula offers landscapes of great beauty and individuality where "zones of respect" and actual national parks have been established. The most famous is found in the southernmost area where the desert peninsula of Ras Mohammed disappears into the Red Sea. Its plateau forms wide sandy beaches while a rich coral reef grows in the underwater depths. Sharm el Sheikh and Naama Bay are situated near the coast on the Gulf of Aqaba, and beyond them lies the spectacular landscape of Uadi Kid, an ancient dry riverbed from the central mountains that touched the coast at Nabq Managed. Vegetation in the Sinai is scarce, with the exception of irrigated areas between the northern coastal plain and the high mountains on the southern peninsula. Only a few perennial bushes survive and grow on the steep slopes of the northern plateau escarpments. Succulents and a few plant species adapted to high concentrations of salt are found on the subdesert terrain of the coastal plains. Most of the population is concentrated along the coast and is exemplified by groups of nomadic Bedouins.

Adrar of the Iforas Desert

Nation:	Mali
Expanse:	49,500 sq km (19,112 sq mi)
Average annual temperature:	6°C–30°C (43°F–86°F)
Rainfall:	<250 mm (<10 in)/year

Above: Stratified sedimentary structures that represent the last southern outposts of the Hoggar Mountains.

Facing page: Encampments of the Kel Ifoghas Tuareg, who are among the least "contaminated" by modernity. Members of this tribe live at the feet of the mountains.

The desert region known as Adrar of the Iforas (also called *Adrar of the Ifoghas*) is the outmost and final southern appendix of the Algerian Hoggar massif. The region is in northwestern Mali on the northern border with Algeria and bordering Niger on the east.

Adrar of the Iforas is a vast plateau whose average height above sea level lies between 600 m (1,970 ft) and 800 m (2,625 ft). The plateau, along with the Hoggar, Aïr and Tibesti Mountains, constitutes the group of four great rock massifs that occupy the central Sahara. With respect to the other three mountain groups, the Adrar is weath-ered enough that its uplands present a softened, rounded morphology and profile that document persistent, intense erosion. The same is true along its deeply carved gorges and wadi. From a purely scenic perspective, the most beautiful ravines are in the Alomamas-Issaouassen area. The many ephemeral water-courses have formed narrow, shadowed valleys where small ponds of water stagnate in abandoned riverbeds. Among the wadi of special scenic interest are those at Es-Souk, where the ruins of a venerable 16th-century Tuareg city called Tadamak may be visited. The Telabit Uadi is shady and humid with florid vegetation and a large number of date palms, while the Uadi Aoukenek is arid and almost entirely occupied by sand.

The abrasion-scoured river valley of the ancient Tilemsi is especially sweeping. The river coursed down from the Adrar Mountains and met the Niger at the city of Gao. The highest peak in the entire massif is Ad Esseli at 890 m (2,920 ft). From its peak one commands a wide plain scattered with the simulacra of ancient mountains. These residual mountains range in height from 300 m (980 ft) to 500 m (1,640 ft).

The Adras region enjoys many oases like Atalaya,

which lies along the road from Tenadjeleline and features numerous and beautiful prehistoric etchings on the rocks that surround it. There is also the In Temssi Oasis with its gardens blooming in the middle of the desert, and the large famous oasis at Telabit where the nearby landscape changes radically. The black of the desert varnish that veneers the rocks replaces the green of the palms in the vicinity of the Aoukenek Well. The well appears to have been visited for thousands of years, considering the different cycles of rock etchings found there that depict animals. Hundreds of local graffiti sites have numerous etchings in *Tifinagh* characters, which is the traditional script of the Tuareg. Groups of Kel Ifoghas Tuaregs live in this region. They are nomads who have kept their traditions more intact than other groups. Insular and not very sociable, they have been feared and respected by other Tuareg tribes for a long time.

Ténéré Desert

Nation:	Niger
Expanse:	400,000 sq km (154,440 sq mi)
Average annual temperature:	24°C–42°C (75°F–108°F)
Rainfall:	25 mm (1 in)/year

Above: A Tuareg follows the crest of a dune at moonrise.

Facing page: The "Great Emptiness," the immense expanse of Algerian sand.

The Ténéré Desert, or "Great Emptiness," is the last southern offshoot of the Sahara. It is a vast expanse gently undulating with dune fields. It lies between the Aïr Mountains to the west, the Tibesti chain on the east, the Djado Plateau and Hoggar Mountains to the north, and the wide depression of the Lake Chad basin on the south.

The famous ergs of the Ténéré—the most spectacular of which is the great Erg of Bilma—are concentrated in the southeastern part of the desert. Flat expanses of *regs* lie to the northwest and are sheltered by mountainous uplands. Isolated outcroppings of sandstone emerge from

these monotonous surfaces. Some of them have characteristic arches due to erosion and subsequent gravitational slippage. These outcrops have become irreplaceable points of reference for travelers. As mentioned earlier, Ténéré means "that which does not exist" in the Tuareg language, and the name mirrors the absolute inhospitality and hostility of this place where one loses perception of time and space.

With the exception of a few oases, the Ténéré is completely free of vegetation. Desertification appears to have started at least 5,000 years ago, and the environment had to have been radically different in that epoch and in the millennia preceding it. Rock drawings of animals and hunt scenes are found at numerous sites; archeological testaments dating back 11,000 years make the Adrar Bous district especially interesting. Many fossilized tree trunks emerge from the sands in the area. Clearly, many watercourses furrowed the Ténéré, as demonstrated by the ancient Tafassesset and Dilia that are hundreds of kilometers long and now dry for most of the year.

An ancient caravan track that crosses the Ténéré dune fields from north to south has been famous since Antiquity. The Bilma, Dibella, Zoo

14

Baba, Chirfa and Timia oases lie along the route. Two nomadic populations who trade in salt and dates, the Kanouri and Tebu, inhabit the Chirfa Oasis. The Kel-oui live at Timia where there is an old fort that features splendid blooming gardens in fierce counterpoint to the nudity and aridity of the sur-

Agadez, home of Tuareg nobility, with Bilma. It was once a huge tree that survived in this arid environment thanks to its 40-m (130-ft) long roots. A Libyan truck driver knocked the tree down during a sandstorm in 1975. A steel structure was erected in its memory and to retain the topographic point

7.7 million hectares (19 million acres) and is the largest protected area in Africa.

rounding landscape. Symbolic of this desert is the Tree of Ténéré, a centenarian acacia recorded on topographic maps and a point of reference for every traveler. It is situated near the caravan route that links the city of

on maps, and the remains of the ancient acacia are now secured in the Niamey Museum. Since 1991, the Ténéré and the Aïr massif have been a nature preserve under the protection of UNESCO; the preserve covers more than

Aïr Massif

Nation:	Niger
Expanse:	41,000 sq km (15,830 sq mi)
Average annual temperature:	28°C (82°F)
Rainfall:	<100 mm (<4 in)/year

The Aïr is a rocky granite massif that rises in central Niger as a continuation of the Algerian Hoggar massif, which it borders on the north. The Aïr's geomorphology presents a classical example of an "inselberg," an "island mountain" typical of desert and subdesert regions, which seems to rise abruptly from the surrounding plains. Extreme thermal fluctuations cause the contraction and expansion of the rocks' outermost surfaces and begin to notch the less-resistant minerals, causing blocks of a variety of sizes to detach and accumulate at the foot of the mountains. Smaller fragments are entrained by the wind and deposited elsewhere. The flanks of the massif are deeply chiseled by canyons called *kory* or *uidian*, which were scoured in remote epochs by the violent erosive action of rushing streams and testify to a climate regime characterized by strong rains that the region once experienced. In the Tezirzek valley near Adrar Chiriet, there are ancient graffiti depicting scenes of daily life that vindicate this proposition. They include hunt scenes and animals typical of the savanna such as giraffes, lions, antelope and ostrich. The course of these rivers is now ephemeral for most of the year and, from the peaks

A granite edifice strongly notched by tectonic and chemical erosion.

of the Aïr, they converge downstream in the Talak alluvial plain. Situated west of the massif, the Talak collects a network of channels that drain toward the Niger River. Cavities within the massif hold thermal springs.

The massif may be observed near the village of Tafedek, which is on the track that links Agadez to Iferouane; the El-mecki tin mines can be seen along the same route. Quite unexpectedly, a good share of Niger's desert lands retains a large number of *gheltas*, the indigenous

Right: An extraordinary giraffe etched in stone, one of many ancient graffiti discovered in the Aïr Massif.

Below: An encampment of the Tuareg Kel, nomads who are becoming ever more rooted.

Facing page: A mountain wadi.

word for small, fresh-water lakes and larger oases like those at Timia, Iferouane and Tassérouakane. Procured from deep aquifers, an abundance of water permits limited grazing and also makes small gardens possible for the cultivation of grain, barley, millet, sorghum, tomatoes, pomegranates, grapefruit and oranges. Acacias and date palms grow spontaneously throughout these areas. The region's population consists

crosses the desert. The climatic and morphological boundary between the true desert and the Sahel lies south of Agadez and is indicated by the passage of isohyet 150 (the line uniting points on a map with equal precipitation expressed in cm/year). Since 1991, the Aïr region, together with the Ténéré Desert, has been a natural reserve protected by UNESCO. At 7.7 million hectares (19 million acres), it is the largest preserve on the African continent (see page 139).

of both nomadic and settled Tuareg Kel. Established primarily at the Timia Oasis, the Kel are primarily engaged in a caravan trade that exploits the Bilma salt flats, located about 600 km (370 mi) from the Aïr massif in the Ténéré Desert. Agadez is the most important populated area in the region and is sited on the mountains' southern slopes. It is more than 900 km (560 mi) from the capital at Niamey and linked to it by a strip of asphalt that

Tibesti Desert

Nation:	Chad
Expanse:	100,000 sq km (38,610 sq mi)
Average annual temperature:	12°C–50°C (54°F–122°F)
Rainfall:	25 mm (1 in)/year

On these pages: Spires, towers and bizarre volcanic structures are the chief characteristic of the Tibesti Mountains.

The Tibesti Mountains (or Massif, as they are often called) occupy an area of 563 sq km (217 sq mi) and belong to the uplands that rise around the central Saharan sandstone plains and plateaus in northwestern Chad between the basin of Lake Chad and the Gulf of Sidra. The mountain chain is almost 480 km (300 mi) long and 280 km (175 mi) wide and follows a northeastern orientation with respect to Niger while meeting Libya at that country's southern border. The Tibesti took form about 70 million years ago in the late Mesozoic period following numerous and repeated volcanic eruptions, which created the great crater and caldera structures that now characterize the landscape. Many of the volcanoes that once existed in the great rock chain are now obliterated by erosion. The most important of the surviving volcanoes include Abeki, Emi Koussi, Tieroko, Tousside, Yega, Oyoye, and Toon—none of which has shown any activity since the Holocene—and Voon, which presently exhibits gentle fumarolic activity. Of these, on the southeast margin of the Tibesti 170 km (106 mi) northwest of the city of Faya is Emi Koussi, also called Mount Koussi, the highest peak in the entire Sahara at 3,415 m (11,205 ft). Rising on the banks of an oasis of the same name and a historical site of trade between the Libyan Fezzan and south-Saharan empires, this volcano is truly exceptional in size. It is a stratovolcano 65 km (40 mi) wide at the base with an immense crater at its summit that descends more than 1,200 m (3,940 ft) into the cone. Trou au Natron is another important volcanic caldera nearly 8 km (5 mi) in diameter and 950 m (3,117 ft) deep. The Tassili Mountains, then, like the Hoggar or Aïr massifs, represent a grand alignment of now-extinct volcanoes, a chain in the shape of an upside-down Y that separates the Fezzan in the north and the Borkou in the south. Surface erosion associated with steep temperature leaps from highs around 50°C (122°F) to lows of −3°C (27°F), and the lithologic variety of lava from dark basalt to redder and lighter-colored rhyolites has created a truly unique landscape full of shape and color. Add to these the ancient wadi or the *enneri*, as they are called in this region, which scoured deep gorges and completely broke down a segment of the massif, attesting once again to a climatic change that rapidly led to the parching of these regions. Rock sites bearing graffiti that date to the era of the great wild fauna also

confirm this thesis. Important archeological sites include Ehi Atron and the Erg of Djourab, with relics dating to the Neolithic, and Gonoa and Oudinger with numerous prehistoric drawings. Of huge importance, finally, is the recent discovery of the

hominid Australopithecus, recovered at Bahr Ek Gazal and christened with the name "Abel."

Ennedi Desert

Nation:	Chad
Expanse:	60,000 sq km (23,170 sq mi)
Average annual temperature:	22°C–39°C (72°F–102°F)
Rainfall:	25 mm (1 in)/year

The Ennedi Desert lies in northeastern Chad, on the border with Libya, which has occupied northern Chad militarily since 1973. Extensive rock plateaus separate it from Sudan. The region's capital is Fada, a typical Saharan oasis town whose urban cluster rose around an old French fort.

Along with the Hoggar and Tassili, the Ennedi represents the last mountain "bastion" of the central Sahara. Ennedi is a sandstone plateau that occupies 60,000 hectares (148,260 acres) of land at an average altitude of about 1,000 m (3,280 ft), with a few dominant mountains as high as 1,450 m (4,757 ft). This desert plateau, whose extension becomes the Libyan Desert, is somewhat triangular in shape. Its sandstone has sustained profound erosion by meteorological agents, especially on the southwestern side, where it has created *pitons*, distinctive and isolated rock crags very similar to the inselbergs found in the Algerian Tassili. The pitons have yardang-like elongated shapes on a northwest and southwest axis created by the prevailing wind that begins to blow south from the Libyan Desert full of sand and crosses the sandy Jourab

Right: The Gueltas, or small oases at the feet of the plateau, permit the growth and sustenance of diversified vegetation.

Facing page: A characteristic "piton" or rocky crag made of sandstone and molded by erosive meteorological agents.

146

17

On these pages: The weathering of tender sandstone and resistant lava creates deep ravines where humid zones are sustained.

region in central Chad. This Saharan landscape must be attributed largely to the erosive agency of surface winds that are tunneled from more stable morphological zones without losing their strength. These zones include the Tibesti Mountains that are largely volcanic in origin and the more fragile sandstone plateau of the Ennedi itself. Between the two mountain masses, the Ennedi is clearly more affected by erosion. Its sandstone is weaker than the Tibesti's resistant volcanic rock, and water has carved a series of wadis and small canyons through nearly 200 km (125 mi) of this sandstone.

The canyon walls are steep and precipitous, but gradu-

ally remodeled and softened by wind to produce towering shapes, spires and natural arches.

The Archei, Tgerkei and Ouni gueltas (oases) are found

at the foot of the plateau. The first warrants particular interest not only for its considerable size and bountiful plant life, but also especially for the beautiful prehistoric drawings found on the rock walls of the natural amphitheater that surrounds it and among its well-hidden ravines. Moreover, the last examplars of the Saharan crocodile, the *Crocodilus niloticus*, survive in spring-fed fresh waters. Evidence of prehistoric life from the lower Paleolithic and ensuing Neolithic is found at many sites hidden within the Ennedi Mountains—10,000-year-old axes, mortars, pestles and arrowheads, but rock drawings and etchings, too. Prehistoric art sites are especially abundant in the Borku region, which is part of the great Saharan complex.

Beyond the gueltas already mentioned, other oases offer lakes in a splendid blue metal-lic color, like Sekhir and Kebir lakes in the Ounianga Oasis surrounded by date palms and by dunes of the great ergs that occupy the Mourdi depression. Finally, there is the Bembéché Plain near the Faya Largeau Oasis where palm groves trail one after another for about 90 km (56 mi), edging the cordons of dunes in the Erg Djourab. This erg occupies the lowest point in Chad, a depression that bottoms out at 160 m (525 ft) above sea level.

Darfur Desert

Nation:	Sudan
Expanse:	440,000 sq km (170,000 sq mi)
Average annual temperature:	8°C–32°C (46°F–90°F)
Rainfall:	<100 m (<4 in)/year

The desert zone occupied by the Darfur Massif is in the western part of Sudan along its border with Chad. Geologically, the massif is an extension of the Basso plateau to the north and the Kapka Mountains to the northwest. It is an imposing highland of volcanic origin that averages 2,200 m (7,218 ft) in height. Jebel Gurgei dominates it at 2,397 m (7,864 ft) above sea level and Jebel Marra, a dark and compact basalt monolith, at 3,088 m (10,130 ft). Ancient volcanic activity is documented by basaltic lava structures that have almost disappeared now. The only traces of any residual heat hidden deep in the terrain are the frequent hot-water springs found on low-altitude hills and at the base of the mountain slopes.

The mass of the Darfur Massif is an important topographic element for the entire region. In fact, it divides the waters of two great hydrographic basins—Lake Chad's to the west and the Nile River's to the east. Many small watercourses also originate in the Darfur. The most important is the Salamat River, which discharges into Lake Chad after running a short distance through Sudanese territory. The connection be-

A satellite image of the Jebel-Marra volcanic area that separates the Nile and Lake Chad basins in the south-southwest. This area also marks the transition between the Libyan Desert in the north and the semiarid savanna in the south.

18

Herbaceous plants and spiny bushes grow in the sandy and rocky soil of the mountains' gentle southern slopes. Fairly frequent rainfall on this side is enough to feed many temporary wadis that run toward the plains south of the mountains. In the north, vegetation is slow growing due to the negative environmental influence of the nearby huge Libyan Desert.

Arab ethnic populations who raise goats and sheep inhabit the northern Darfur and its deep ravines while other Arab groups live together with the Fur in the south. These latter, along with small groups of *Meja Zaghawa*, and *Daju* nomads, cultivate small plots of cereal grains, millet, sesame and fruit. The products of these craftsmen are sold in the markets of Nyala, the region's capital and point of departure for exploring the surrounding mountains.

151

tween the Darfur Mountains and the Nile's alluvial plain slopes gently and gradually through a part of the river bound by the Third and Fourth cataracts and is a vast lowland plain called the Dongala.

Nubian Desert

Nation:	Sudan
Expanse:	407,000 sq km (162,800 sq mi)
Average annual temperature:	10°C–45°C (50°F–113°F)
Rainfall:	15–40 mm (0.6–1.6 in)/year

The Nubian Desert, whose original name is *As-Sahara' An-nubya*, occupies the northeastern region of Sudan, a territory bound on the west and south by the Nile River, which flows over topographic leaps that create rapids and waterfalls through this stretch. The desert is bound on the east by the coast of the Red Sea, while it marks the border with Egypt in the north. The region's rugged morphology offers a highly varied landscape characterized by greatly eroded rock areas and rough rock plains, alternating with systems of sand dunes and sheets of dust. Part of the region is deeply and densely carved by numerous wadis. Their ephemeral streams head for the Nile but never reach it before disappearing into the desert itself. The plains undulate gently on a slope, which increases little by little as it approaches the Coral Sea, where it forms a steep rock plateau embodied by the Red Sea Hills. Despite their name, these hills are true mountains whose highest peak is the Jabal Erba at 2,217 m (7,274 ft).

Sand and dust storms locally known as *haboob* are characteristic of the Sudanese region. They occur fairly often during the hottest and driest months preceding the humid season. Rains are minimal, however, even if there were surely more rainy episodes and eras in the past, enough to have created permanent pools of water, which no longer exist and have left the field to flat, dry playas. One of the most famous and interesting playas is near Nabta, where the remains of an important Megalithic site

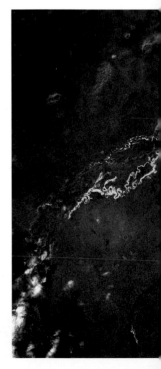

A satellite image shows a semiarid zone in the desert southwest of the course of the White Nile. It shows many ephemeral watercourses (wadi) that skim the 1,065-m (3,494-ft) high Dair massif. Groups of linear dunes are clearly visible at the top right of the photograph.

are found, including a circle of flat stone slabs similar to tombs and containing the bones of animals that were probably domesticated. Around the "tombs" there is a row of megaliths made of sandstone from a rock outcrop about a kilometer from the site. Roughly crafted, the stones reach the extraordinary height of 10 m (33 ft) and are arranged in two orthogonal alignments, one north-south and the other east-west. Indigenous peoples frequented the site to conduct religious rites that were somehow linked to specific astronomical events like the summer solstice. Nomad populations and permanent communities both inhabited the Nabta region until about 4,800 years ago when the inexorable processes of desertification caused the entire region to be abandoned. This part of Sudan historically belonged to the ancient region of Nubia, which comprised the territories between the River Nile's First Cataract in Egypt and Khartoum in Sudan. Under Egyptian dominion, which lasted almost four centuries, the territory was known as the *land of the Kusti* and was exploited for its reserves of spices, ivory and lumber and its traffic in slaves. The entire area was Christianized during Justinian's reign. The remains of ancient churches in many Nubian cities testify to these events, and the most famous are surely those found in ruins at Faras. The Nubian Desert is now a small and isolated piece of the Sahara. Its few remaining villages are found along the banks of the Nile and on the banks of intermittent wadi. Its minimal economic activities are concentrated here but now reduced to the cultivation of dates, rare fruit trees, a few plots of wheat and sheep raising.

Sahel

Nation:	Sub-Saharan Africa
Expanse:	5,000,000 sq km (1,930,500 sq mi)
Average annual temperature:	25°C (77°F)
Rainfall:	150–250 mm (6–10 in)/year

The Sahel is a very large region that occupies part of western and central-northern Africa. It extends from the Atlantic Ocean east across

cal environment is semiarid steppe characterized by slow-growing grasses and large herbaceous perennial plants, spiny bushes, and acacia and baobab trees. Its vegetation is sufficient to produce food for a prudent number of livestock including camels, oxen and other grazing species.

The spiny scrub was once underbrush and now covers

Above: Cattle raising and imprudent exploitation of the land have contributed to the increased desertification of this territory.

Right: The last spurs of the desert graze the waters of the Atlantic Ocean.

Senegal, southern Mauritania, the large strip of the Niger River in Mali, Burkina Faso, southern Niger, northwestern Nigeria, central and southern Chad and finally, Sudan. It is an important transitional area that ideally divides the Sahara Desert on the north from the band of humid savanna to the south. The climate is dry for at least eight months a year and the rain, limited to a brief season between June and August, dumps an average of 100–200 mm (4–8 in) of water each year. This territory's typi-

the terrain inconsistently, presaging an irreversible trend toward the complete desertification of the Sahel.

Beyond climatic factors, this desertification is also linked to mistaken agricultural and grazing practices that have definitively and irreversibly impoverished the terrain. This tendency has been obvious since the second half of the 20th century when the Sahel was overwhelmed by escalating soil erosion concurrent with the farmers have used and continue to use a huge quantity of trees and bushes as firewood or for agricultural purposes. A direct consequence of this excessive exploitation of soil resources is rainwater that flows across the ground without penetrating it and, with the wind, erodes the topmost part of the terrain— the only layer that possesses flooding along the Niger and Senegal rivers, which permits the cultivation of millet and peanuts.

But the Sahel is a fragile place at the mercy of climatic events. In 1970, after almost three years of dryness, the territory was virtually denuded of crops and more than half its livestock was lost. After brief rains in 1972, it was again completely deprived of precipitation and the Sahara Desert advanced more than 100 km (60 mi) south in the course of the following year. The direct loss of human life caused by this climate situation was estimated at about 100,000 people in 1973. Governmental intervention aimed at the reforestation of small areas has not been adequate to stop the desert's inexorable advance and it is occupying the steppes of the Sahel ever more rapidly.

exceptional growth of population and herds that reduce the herbaceous cover.

Both city dwellers and attributes of fertility—leaving an arid and absolutely unproductive land. There are some exceptions, such as occasional

Mauritania Desert

Nation:	Mauritania
Expanse:	720,000 sq km (278,000 sq mi)
Average annual temperature:	38°C (100°F) in the summer
Rainfall:	<80 mm (<3.2 in)/year

Above: A stone stele marks the site of an ancient burial.

Facing page, above: Small interdune oases where wells are dug to furnish water.

The Mauritania Desert is the western extremity of the Saharan Desert. Its expanse is confined on the south and east by Mali, on the northeast by Algeria, on the west by the Atlantic Ocean and on the northwest by the Western Sahara. It contains two vast plateaus that reach remarkable heights: the Adrar on the north and the Tangat on the south. These are surrounded by the extensive Erg Al-Djouf dune fields on the north and the Erg Aoukar on the south. The first is characterized by long cordons of dunes that move along the coast in a northeast-southwest orientation. The Adrar plateau is located about 500 km (310 mi) from the capital of Nouakchott, and is composed of a vast tableland cut by deep canyons and featuring small oases at their base. Mount Kedjetb al-jill, at 915 m (3,000 ft) the highest peak in Mauritania, is found here. The ghost city of Oudane sits near the eastern edges of the plateau. Once a rich center for packaging gold and salt, it has fallen victim to the inescapable march of desertification.

A few miles away in a place called Guelb er Richat, there is a tectonic dome almost 40 km (25 mi) in diameter caused by sedimentary rocks cut off by erosion. The rock layers form a concentric "bull's-eye" structure. A lake that once sat in this depression no longer exists, but it left a thin layer of shells on the surface. The Tagant—meaning "forest"— Plateau lies about 300 km (186 mi) southeast of Nouakchott. Its sheer walls with 600-m (2,000-ft) drop-offs face the sandy el-Djouf and Aoukar basins. Underground springs in many canyons permit the cultivation of fertile fields before draining 300 km (186 mi) south to the Senegal River. Stone artifacts that date to the Neolithic period have been found at the Aratane Oasis in the Tagant interior.

The geologic history of Mauritania and its desert starts in the distant Precambrian, about 600 million years ago. At the time, a series of grand-scale tectonic movements alternately governed the uplift of the North African gneisses and granites and its sinking to the point of submersion in the Cambrian and Silurian seas. Sedimentary formations—generally sandstone—were often deposited on the floor of these basins. The region's final emergence

started at the end of the Carboniferous Period 280 million years ago. As these ancient seabeds heaved up, they maintained their horizontal position, giving origin to vast plateaus. In 1976, a part of the desert with its fragile ecosystem was incorporated into the Banc d'Arguin National Park about 200 km (125 mi) from the capital. The area covers more than a million hectares (2,471,000 acres) and accommodates several environments and morphologies, from sand dunes that are home to the gazelle to coastal mangroves where birds take refuge and fish proliferate. The park came under UNESCO protection in 1989.

Somalia Desert

Nation:	Somalia
Expanse:	320,000 sq km (124,000 sq mi)
Average annual temperature:	40°C (104°F)
Rainfall:	<50 mm (<2 in)/year

Above: A herd of camels moves slowly across the plateau to reach the rare oases.

Facing page, above: The site of an ancient sepulchre on the Mudug plateau; below, the Mudug tableland ranges in height from 600 m (1,970 ft) to 800 m (2,625 ft); its rock is volcanic in origin, alternating with granite.

A torrid climate and scarce precipitation are typical conditions in northern Somalia, a territory generically referred to as the "Somalian plateau" and occupied by a series of mountain chains and plateaus of volcanic origin. Djibouti borders the area on the west, Ethiopia on the south, the Gulf of Aden on the north and the coast of the Indian Ocean on the east.

The region constitutes the direct eastern continuation of the Ethiopian plateau, and breaks down into a vast tableland called the Mudug to the southeast. Between 600–800 m (1,970–2,625 ft) above sea level, the Mudug gently loses altitude toward the south and grades into the low Indian Ocean coast that is rimmed by dunes and, locally, by coastal lagoons. The Cal Madow chain runs parallel to the northern coast where its mountains descend steeply through a series of terraces toward a limited strip of coastal land. Surud Ad, the highest peak in Somalia at 2,408 m (7,900 ft), dominates the mountain chain. The Ogo Plain lies along the northern border with Ethiopia. It is the continuation of the Hawd Plateau and is also called Galgodon. The Nugaaleed Valley lies in an east-southwest direction at the foot of the chain and is the remnant ravine of an ancient stream, which is now dry most of the year. The Benadir is the only broad, fertile alluvial plain in all of Somalia and as such is intensely cultivated. It lies in the south between the Uebi and Giuba rivers, the only rivers that carry a year-round flow of water. The Benadir's unusually regular morphology is interrupted here and there by inselbergs or isolated granite outcrops.

The Somalia Desert's climate is equatorial in the south and arid in the north and inland with less than 150 mm (6 in) of precipitation a year. Rains are relatively abundant only in the south and may total 279 mm (11 in) a year. Only a few plant species can live in the plateau's extremely arid environment. There is an unregulated harvest of its few trees and bushes for firewood or to make more land available for agricultural use, but this only aggravates the environmental degradation. Most of the vegetation is concentrated along seasonal streams and is mostly evergreen shrubs. Higher up the plateau—at 1,500 m (4,922 ft)—acacias dominate the landscape where the water table is fairly shallow along with juniper and cedar bushes protected by the Somalian government and termite colonies. These colonies look like isolated muddy burial mounds—or tumulus—built with a paste of predigested wood, sand and a mix of secretions that make the internally hollow structure extremely sturdy.

Danakil Desert

Nation:	Somalia
Expanse:	266,000 sq km (102,700 sq mi)
Average annual temperature:	10°C–27°C (50°F–81°F)
Rainfall:	<25 mm (<1 in)/year

Stunted vegetation ekes out survival on the banks of Lake Afrera, also known as Lake Giulietti.

The Danakil Desert includes the coastal band of Eritrea with an average width of 150 km (95 mi) and the vast plateau that occupies the northern part that is called "Upper Africa," corresponding to ancient Abyssinia. It is among the world's most inhospitable land with daily summer temperatures exceeding 50°C (122°F) in the shade, but then there seems to be no shade in the Danakil. The territory's current geologic structure is the result of tectonic movements that led to the separation of Africa from southwestern Asia and thus to the formation of the Rift Valley and, eventually, to the Danakil Depression. These great geomorphologic features divide two highly distinct territories with the Ethiopian plateau on one side and, on the other, plateaus that incline toward the low tablelands and outermost plains of Somalia. The territory abuts a segment of the Nubian-Arabic shield composed primarily of schist, gneiss and granite. The shield was transformed by an archeozoic orogeny, which governed the formation of the mountain chains that were subject to continual and intense erosion during the Paleozoic. A slow sinking process began at the beginning of the Mesozoic, when a vast plain had already replaced the moun-

tains. As the sea gradually invaded Somalia, Danakil, and part of the Ethiopian plateau, clastic marine deposition led to the formation of sandstone.

Limestone deposits were created farther offshore where deeper ocean conditions had been established. Gradually, the leveling process was arrested and replaced by a slow uplift and a corresponding retreat of the sea; allowing additional shallow-water sandstone to be laid down atop the limestone as the region re-emerged. Early in the Ceno-

zoic, great tectonic upheavals fragmented the crystalline shield. Lava flowed out of these deep gaps and spread up to 3,000 m (9,840 ft) thick across the sedimentary surfaces. In the meantime, with the opening of the Red Sea, the Danakil Depression was invaded by the sea and the Rift valleys began to separate from it. The relatively complex character of this area derives from ongoing tectonic and volcanic activity, various climate cycles and discontinuous erosion. Among other phenomena, lava flows dammed up several valleys, disrupting drainage patterns and creating lakes like the Tana while isolating the Dancalo Gulf from the ocean. Intense erosion instigated by

Right: Lava tablelands and tectonic depressions characterize the Danakil landscape.

Below right: Thick saline crusts on the banks of Lake Afrera.

Below: The intense evaporation of Lake Afrera by fierce heat results in huge salt deposits.

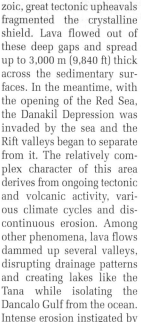

Afrera. Some of the world's most interesting active volcanoes such as the Marahó rise from this desolate plain. Erta Ale is one of the few that still has a lake of lava bubbling perennially in its crater, while poisonous gases and vapors escape from the Afdera volcano. Active volcanism combined with very arid and inhospitable conditions render the Danakil Desert unique and extreme.

tectonic uplift opened deep gullies in the tabular complex of volcanic rock and sandstone. Nowadays, these canyons are important morphological elements because of their location at the base of the country's regional divide. Thus, the Abbai Valley (Blue Nile) divides Goggiam from Scioa and Uoleggà; the Tacazzè (Tekeze) separates the Tigré from Amhara. Tabular elements known as *ambe* rise among the canyons and represent the characteristic morphology of the Ethiopian highlands whose highest peaks are in the Semièn Mountains, culminating in the 4,550-m (14,928-ft) Ras Dashen.

Many other summits in the region exceed 4,000 m (13,120 ft), such as Abuna Josef (4,190 m/13,747 ft), Guna (4,231m/13,882 ft), and the Mangestu Mountains (4,100m/13,452 ft). The arid and desolate Danakil covers 100,000 sq km (38,610 sq mi) and is bound on the east by the so-called Dancal Alps, a mountain system with a tabular structure and a few volcanic cones that reach their maximum height in Mount Ramlo (2,130 m/6,980 ft).

The flatlands host a plinth of salt up to 800 m (2,625 ft) thick in the Salt Plain. It bears distinctive cryptodepressions, including Lake Asale and Lake Giulietti, which fall below sea level by 116 m (380 ft) and 80 m (260 ft), respectively. Lake Giulietti is better known as the

Chalbi Desert

Nation:	Kenya
Expanse:	180,000 sq km (70,000 sq mi)
Average annual temperature:	40°C (104°F)
Rainfall:	<150 mm (<6 in)/year

Chalbi itself. The plateau is covered with a luxuriant forest covering 2,000 sq km (772 sq mi) and sheltering numerous plant and animal species, surrounded by a hostile, arid

On these pages: Concentrated near Lake Rudolph, the biggest lake in Kenya, Turkana and Samburu ethnic tribes are in conflict.

Located in northern Kenya along the Ethiopian border, the Chalbi Desert is part of a great variety of ecosystems found in Kenyan territory including rainforests, prairies, savannas and desert steppes. The true desert stretches east from the Rift Valley, the great tectonic fracture that stretches thousands of miles north and south, dividing all of eastern Africa. The Marsabit Plateau, an ancient extinct volcano, dominates its sand expanse; it is almost 1,700 m (5,575 ft) high and located south of the

and scorching-hot territory where wind velocities range to 100 km (60 mi) and more per hour. The oldest national park in Kenya, established in 1900, takes its name from this desert. Lions, giraffes, leopards, hyenas, spiral-horned kudu and elephants live in the interior. The symbol of Kenya, the famous elephant Hamed, now stuffed and kept in the National Museum at Nairobi, lived here.

The Marsabit Plateau is the only place in northern Kenya that enjoys abundant

164

fossil footprints of ancient prehistoric people can be followed along its eastern shore. A few exceptional discoveries that proved useful in reconstructing the evolution of the genus *Homo* were made in the 1970s near Koobi Fora. In 1971, Richard Leakey came across a complete male cranium here dating back 1.6 million years and attributed to *Homo Erectus*. This artifact is now kept in the National Museum of Kenya in Nairobi along with Hamed.

precipitation, which comes in two seasons, from March to June and October to December. During these intervals, small salt lakes called *gof* form on the floor of many small depressions found across the entire area. Lake Turkana—the former Lake Rudolph—forms the western edge of the desert. It covers 6,405 sq km (2,473 sq mi) and is the largest in Kenya. Brimming with fish, its banks support settlements of the *Turkana* and *Samburu* tribes who are often in conflict.

The lake's paleo-anthropologic reliquary is of considerable merit. Extraordinary

The Kalahari

Nation:	Botswana
Expanse:	260,000 sq km (100,390 sq mi)
Average annual temperature:	summer 40°C (104°F); winter 0°C (32°F)
Rainfall:	120–700 mm (4.75–27 in)/year

The Kalahari covers an area of 260,000 sq km (100,390 sq mi) in Botswana, eastern Namibia and northern South Africa. Its mixed environ-

hundred square meters and a few square miles. The pans capture drainage and rainwater and may hold them for several months, attracting both people and animals. However, the highly saline water is not nearly as important as the vegetation that grows around the pans.

The Kalahari spreads from the Orange River in South Africa to the southern Congo.

Right: The arid and clayey surface is hardened by the sun and displays the typical structure of desiccated polygons.

Facing page: The "pan of the Baobabs" is a water-rich depression where baobab trees have been growing for hundreds of years.

ments include sand dunes, savanna and prairies. Rainfall measures between 120 mm (4.75 in) in the southwest and 700 mm (27 in) in the northwest. As the inland extension of the Namib Desert, the Kalahari is considered a coastal desert with a semi-arid environment.

Pans are typical of the Kalahari. These are rounded, hollowed out areas with beds of hardened gray clay whose dimensions vary by a few

The largest inland delta—the delta of the Okavango River—loses its way among the Kalahari's desert sands. The desert's vegetation includes grasslands, bushes and palms and is more abundant around the pans. Its highly diversified fauna includes lions, leopards, cheetahs, hyenas, jackals, elephants, zebras, rhinoceroses, giraffes, hippopotamuses, foxes and various species of antelope. Bird populations are estimated to

include about 400 different species.

Parks and reserves protect some of the Kalahari ecosystem. Created in 1961, the Central Kalahari Game Reserve is a 52,000 sq km (20,077 sq mi) reserve established for indigenous bushmen known locally as *San* or *Kung*. It is thought that these peoples have lived in the Kalahari for nearly 30,000 years. Totaling 36,000 sq km (13,900 sq mi), Gemsbok National Park and the Kalahari Gemsbok National Park are contiguous to allow migration of their large animal populations. They were created in 1931 to reduce

poaching and are home to an immense area covered by dunes of red sands. Other parks are the Game Reserve of Mabuasehube, set up to preserve the ecosystem of the pans; the Madikwe Game Reserve, established to save the black rhinoceros; and the Tswalu Private Game Reserve.

The Kalahari accommodates various species in danger of extinction; among the most famous is the cheetah. At the beginning of the 20th century, the cheetah population was about 100,000; it was 25,000 in 1980. Now, only about 10,000 cheetahs survive in the wild. Many

preservationist and research groups are actively engaged in habitat conservation and educating local inhabitants. Even the black rhinoceros has suffered sharp declines in its population despite having its own preserve. The 75,000 rhinoceros alive in 1970 have been reduced to about 2,500 today. Illegal hunting is the primary cause of this reduction to near-extinction, but it must also be remembered that the entire animal population of Botswana has been reduced from its historic maximum by 96% in the last 10 years.

Local herders overgrazing

Above: A termite colony that has attained a height of 2 m (6.5 ft).

Facing page: A centenarian baobab in the middle of a Botswanan pan.

their livestock cause the most damage to ecosystems. The effect of the increasing number of sheep, goats and karakul (a kind of sheep) has been to denude the vegetative cover and facilitate the advance of desert sands. Competition between man and animals for the same resources has caused a drastic reduction of available resources while ongoing climate variations reduce the quantity of precipitation.

Native populations have also fared poorly. Persecution and epidemics have reduced the Kung population to 85,000 people, spread around isolated areas of Botswana, Angola and Namibia. The Kalahari environment they live in is truly hostile and every other population has avoided it, but the Kung have adapted to survival in this milieu. Their villages include 10–30 persons and are semi-permanent

leader but govern by group consensus. Disputes are resolved after lengthy discussions in which each person may express an opinion. During the rainy season, when water and food are abundant, the bands move around to visit relatives. During the dry winter months, the Kung remain near springs of water.

inasmuch as they are dependent upon water supplies. A fire burns constantly in front of every hut. The Kung are a hunter-gatherer people who possess a remarkable knowledge of the edibility of food and the curative or toxic properties of various plant species available. Wildlife is sparse in their territory so the hunters often have to cover huge distances. The bands do not have a true

Namib Desert

Nation:	Namibia
Expanse:	94,000 sq km (36,300 sq mi)
Average annual temperature:	12°C–35°C (54°F–95°F)
Rainfall:	<50 mm (<2 in)/year

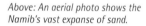

Above: An aerial photo shows the Namib's vast expanse of sand.

Facing page: Clusters of oryx graze on the high, red sand dunes.

The Namib Desert occupies the Atlantic coast of southwestern Africa for at least 2,080 km (1,294 mi) from the city of Namibe in Angola (where it is called the Mocamedes Desert) and south across Namibia to the Olifants River in South Africa's Cape Province. Its desert terrain reaches almost 160 km (100 mi) inland, uniting with the Kalahari Desert. The Namib is bordered on the east by the Namibian region of the continental scarp that runs all around southern Africa. Canyons and rock walls characterize the scarp and landslides are frequent. Beyond the scarp, the African plateau rises 1,000 m (3,280 ft) and more in height; its western border is the Atlantic coast. The Kuiseb River, which disgorges into the Atlantic near Walvas Bay, cuts the Namib in half. To the south, there is a vast sand sea with three types of dunes. First, there are crescent dunes typical of coastal deserts with constant

prevailing winds from the ocean, then parallel linear dunes 16–32 km (10–20 mi) long and as high as 300 m (980 ft), separated by wide hollows in the vicinity of an ancient saline basin called Suossus Vlei. Finally, there are star dunes 300–350 m (980–1,150 ft) high with long radial arms that reflect the rather chaotic prevailing winds in this part of the desert. Underlying them are ancient cobbled terraces where the world's largest deposit of gem-quality diamonds is found. The Namib's sand is derived primarily from fluvial sediments transported by the few perennial rivers that empty into the ocean, but some is clearly marine in origin.

The landscape changes completely north of the dry Kuiseb River bed, a barrier to the eolian transport of sand. The landscape becomes a platform of igneous and stony gray metamorphic rock

Above: A thick crust of salt covers a topographic depression.

Facing page: The sand sea extends for hundreds of kilometers; giant barchans exceed 350 m (1,148 ft) in height.

(schist, quartzite, granite) that is as smooth save for isolated inselberg rock formations.

Temperatures along the coast float between 10°C–16°C (50°F–61°F) with a minimal thermal excursion between day and night and summer and winter. Farther inland, they range from 31°C–38°C (88°F–100°F). Exceptionally warm temperatures prevail only a few days a year when winds blowing from the east carry higher temperatures, dry air, and clouds of sand and dust.

Snow falls on the highest southern mountains on rare occasions, and below-freezing temperatures are recorded along the desert's inland extremities.

The Namib is the driest African desert. Rain falls only a few days a year in the spring or in occasional short storms. The average annual precipitation is 50 mm (2 in) farthest inland and decreases to just 13 mm (0.5 in) as it moves west toward the coast, where practically the only form of humidity is nocturnal dew or the fog produced by the cold Benguela current that flows offshore. This measure of humidity contributes about 4 mm (0.15 in) to the annual average precipitation. With its paucity of rain, the Namib has a barely developed surface hydrographic network. The only permanent rivers are the Orange and the Cunene, and any others normally contain water for only a few days a year and only in exceptionally rainy years. The larger northern rivers reach the ocean, but those between the Kuiseb and the Orange all terminate in salt lakes or muddy plains among the dunes.

Most of the Namib is virtually inhospitable. The few lands that can be cultivated are limited to the flood plains and to the terracing of the most important rivers. Though scarce, local vegetation is rather varied. Typical of this region, the *Welwitscia mirabilis* unfolds two enormous leaves in opposite directions along the ground. The fauna is also surprisingly varied and rich, and more so than in any other sandy desert. For this reason, national parks like that on the Skeleton Coast or Naukluft Park protect 15% of Namib territory. The latter is the largest park, with 50,000 sq km (19,300 sq mi), and is known as the "Living Desert." It is a plateau that rises over a desert plain and is heavily etched by steep and deep ravines. The highest dunes in the world are "confined" there, rising over 400 m (1,300 ft). Numerous animal species find an ideal habitat there, including mountain zebras, baboons, oryx and cheetahs. Among the most characteristic and interesting places in this desert are the houses of Kolmannskop, a ghost city near Luderitz; Swakopmund's lunar-looking rock formations to the west; and Sandwich Harbour, where imposing dunes reach the ocean.

Andalusian Desert

Nation:	Spain
Expanse:	266,000 sq km (102,700 sq mi)
Average annual temperature:	10°C – 27°C (50°F– 81°F)
Rainfall:	<100 mm (<4 in)/year

Above: A small accumulation of sand imprisoned by a bush may give life to a stabilized dune.

Right: High dunes of extremely fine sand in the National Park.

Far right: Desert dunes migrate to the ocean shores.

Spain is surely one of the wildest regions in Europe. Its land, in places, appears to have suffered incessant and fierce soil erosion. A great rock tableland occupies much of the country and more than half of its entire territory. Its morphology is the outcome of rapid desertification due in part to human activity that started a few millennia ago. In the fourth century B.C., the great Greek philosopher Plato described the phenomenon of desiccation with regret in an unfinished manuscript: "Erosion devours the mountain soil without pause. The fine rich soil slides into valleys and the sea swallows it. . . . It washes away and only a skeleton of the landscape is left."

In the generally arid landscape of the Iberian Peninsula, the Almería region is the most eroded and desolate part of the continent and is considered a true desert on a par with its more famous North African counterparts. It lies between the Sierra Morena in the north and the Sierra Nevada on the east and occupies the extreme southeast corner of Spain. A little north of the Cabo de Gata and as far as the slopes of the Sierra de los Filares, the ground is barren of any kind of soil, and only its clay substrate dating to the Miocene is visible. Nor is there any vegetation whatsoever on the deeply eroded Tabernas region, a classic rain shadow downwind of the Sierra Nevada. Temperatures there can reach 45°C (113°F), and precipitation does not exceed 110 mm (4.25 in) per year. Concentrated in rare but violent downpours, the Tabernas rains further impoverish the terrain with their sheet-washing action. The

final result is a landscape not found anywhere else in Europe, composed of a long series of trenches sometimes covered by a thin layer of sand, free of vegetation, and monotonously repeated for kilometers on end.

The desert extends to the south, almost as though it wished to unite with Africa. Southwest of Antequera near Torcal National Park, wind and meteorological action sculpt peculiar rock structures in bizarre shapes. Toward the Garanta del Chorro at the edge of these stone pillars, a huge expanse of pine trees was recently planted and is protected in an effort to check soil erosion. There, a second small desert, called the Arenas Gordas, has a large dune field 60 km (37 mi) long and more than 14 km (9 mi) wide. It is near the Cota Donana area at the mouth of the Rio Fuadalquivir, which empties into the Gulf of Cadiz. Near El Asperillo the dune fields reach a height of 100 m (330 ft) and are thus the tallest dunes in Europe. The ultimate source of the sand is the Sierra Morena and Sierra Nevada; eroded rock debris is carried to the sea, entrained by coastal currents and winds, and heaped up inland. This process of erosion and deposition continues and creates the big field of mobile dunes that advances inland at a pace of 9 m (30 ft) a year. The boundaries of this desert in miniature, which is characterized by the great variety and rapid evolution of its contours, are the ocean on one side and the rock promontory at Sanlucar de Barrameda on the other.

MIDDLE EAST

Both landscape morphology and aridity vary considerably from one place to another in the Middle East. Great mountain chains line one side of the area, including the Zagros and the mountains of Elburz in Iran, the Tauri in Turkey, the Asir Mountains of Arabia and the Jabal al Akhdar in Oman. On the other side, there are large inland plains and the Arabian plateau with its two great sand seas of An Nafud and Rub' al Khalim and ample intermountain basins with *kavirs* (salt plains) in Iran.

This diversity is largely caused by the region's tectonic history. Much of Arabia represents the remains of the Gondwana continent, while its mountain chains are associated with the interaction of three great plates: the African, Eurasian and Arabian.

Environmental changes that occurred during the Tertiary and the Pleistocene are also responsible for the region's variety of landscapes. Humid conditions prevailed during the last part of the Tertiary in the region, attested to by the presence of deeply weathered, 3.5-million-year-old basalt flows supporting a thick laterite soil cover. By contrast, younger flows that date to the Lower Pleistocene (about 1.2 million years ago) show no alteration at all.

During the Quaternary, there was no long-lasting humidity despite some evidence of rainy periods, for example, in the Dead Sea depression. Aridity was probably dominant for most of the Pleistocene. In Saudi Arabia, alluvial deposits from the Upper Pleistocene, calcretes and lacustrine deposits clearly define a rainy episode between 24,000 and 33,000 years ago, with a more intense humid phase between 26,000 and 28,000 years ago. There were large lakes in the Rub' al Khali at this time. Eolian activity prevailed between 10,000 and 19,000 years ago in Arabia, and dunes formed on the floor of the Arab Gulf when its sea level was lower. About 9,000 years ago or slightly later, rainy conditions returned, causing the formation of lacustrine deposits in the Rub' al Khali, and limestone soils and alluvial deposits along the principal wadis. These environmental conditions lasted until about 5,000 years ago. In the Konya Basin, in one of the most arid zones on Turkey's Anatolia plateau, a series of ancient lake shorelines have allowed researchers to reconstruct the Near East's climatic and hydrologic variations. The lake shows three phases of maximum extension: Konya I, older than 30,000 years (radiocarbon dating sets the maximum at 35,000–40,000 years ago); Konya II, between 17,000 and 23,000 years ago; and Konya III, from 11,000 and 12,000 years ago. Konya II coincided with the last glaciation of the northern hemisphere. The lake disappeared 17,000 years ago, before the retreat of the Fennoscandia and Laurentian ice caps. The lacustrine phase at the beginning of the Holocene recorded in Saudi Arabia is not represented in the Konya sequence. According to some authors, this

is true because the paleolakes in Iran and Anatolia were caused by decreased evaporation due to reduced temperatures rather than by increased precipitation.

Recent studies have found evidence of remarkable variations in environmental conditions from place to place in the Middle East, causing revisions of climate models made prior to the 1980s. These differences probably depend on the relative importance in different times and places of major climatic systems like the summer monsoon, western cyclonic winds, and masses of cold Eurasian air. For example, during the last glaciation, the Mediterranean's northern coasts were subject to the influence of masses of cold and dry air generated by the Eurasian ice caps. Its southern coasts took in precipitation from the western cyclonic systems that were diverted and forced between the cold north-ern air and the more or less fixed subtropical belts of high pressure. Southern Arabia was influenced by monsoons and therefore its climatic tendencies show a certain similarity to those observed in Eastern Africa and in the Thar.

Arabian Desert

Nation:	Saudi Arabia
Expanse:	2,300,000 sq km (888,030 sq mi)
Average annual temperature:	maximums exceed 54°C (129°F)
Rainfall:	100 mm (4 in)/year

Above: A broad surface covered by flint deposits on the Jordanian border.

Right: Granite walls eroded by meteorological agents in Jordanian territory.

The Arabian Desert is an immense region that occupies 90% of the entire Arabian Peninsula. The Syrian Desert and the Jordan Plateau, which rises more than 1,000 m (3,280 ft) above the Dead Sea, border it on the north. About 70 km (40 mi) from Aqaba in Jordan, a unique place called Uadi Rum stretches across solid desert. It is a succession of deep valleys, imposing sandstone outcrops in grotesque shapes and colors, and deep canyons with walls more than 500 m (1,640 ft) high that conceal old archeological sites and prehistoric art.

The Arabian Desert meets the Iraqi plains of the Tigris and the Euphrates to the northwest, reaching as far as the Persian Gulf and the Gulf of Oman. To the southeast, it terminates at the Arab Sea and the Gulf of Aden and touches the shores of the Red Sea on the east. Most of the Arabian Desert lies within the modern kingdom of Saudi Arabia and partly in Yemen, Oman, the United Arab Emirates, Qatar, Kuwait, a small part of Iraq and Jordan. It reaches its maximum length of 2,600 km (1,617 mi) in a northwest-southeast direction.

Geologically, the Arabian Peninsula is the continuation of the Saharan shield. The shield experienced rifting

where the Red Sea sits today and was then lifted along its entire western and southern borders, giving rise to a series of volcanic mountains bordering the Red Sea and the Gulf of Aden. The highest peak in this jagged mountainous region is 3,730-m (12,238-ft) high Mount An-Nabi Shu'ayb on the Yemeni coast.

Northeast toward the heart of the Arabian Peninsula and the Persian Gulf, the elevations merge into an utterly vast interior plateau that, in most places, lies below 1,000 m (3,280 ft). Here, igneous and metamorphic rocks grade into younger sedimentary rocks. The plateau can be sub-

Above: A siliciferous zone free of plant life on the Jordanian border.

Facing page: A succession of sculpted sandstone and granite at the edge of the Uadi Rum.

divided into three principal areas: the Nafud to the northwest; the Rub' al Khali on the southeast; and wide basins of sand dunes separated by a central mountainous zone, the Neged. A series of volcanic areas separate the great sandy deserts in the Nafud and the Rub' al Khali from the western coastal mountains. Recurrent plateaus and the wide, rocky, and siliceous plains that border its two great desert basins on the east characterize this desert, as does the sebkha. A typical landform in the eastern Arabian Desert, the sebkha is a plain of salt crusts inclined toward the sea, created when marine waters are entrapped and then evaporated from inland areas. The Sebkha

Matti plain is the largest exposed sebkha and some of it consists of soft, wet, saline mud over a hard crust about 90 cm (35 in) deep. The surface of the sebkha is not directly analogous to quicksand, but it is nonetheless a danger to the inexperienced traveler, who may fail to recognize its nature quickly enough to avoid sinking into the mud.

The Neged region is a mountainous zone of low limestone crests that rises toward the center of the peninsula between the two Nafud and Rub' al Khali Deserts. Two long parallel arcs of dunes link the two sand basins to one another. The Ad-Dahna is the larger arc and it surrounds the

Neged massif on the east. In the center area, sand accumulates in enormous quantities in the two main basins, covering the plains and lower hills. Regular, mobile dunes that may grow to 90 m (295 ft) and resemble a hoof characterize the Nafud. The Rub' al-Khali seems far more varied, however, with dunes, crests and gigantic complexes that develop in successive belts that are linked to the advance and the retreat of the ocean. Dunes in this basin take the shape of barchans, seifs, domes, pyramids, crescents and the letter S.

Occasional torrential rains that refill the region's ever-arid streams are fundamental in shaping the landscape and finally prove far more brutal than the erosive force of the wind. The huge regional watershed of mountainous crests along the coasts of the Red Sea and the Gulf of Aden determine the desert's hydrographic network. A series of short streams empty into the sea after cutting deep canyons while longer streams on the opposing and more gradual inland slopes get lost on the huge desert plateau. Beyond these coastal systems, there are other dry or intermittent basins inland. Rainfall in the Arabian Desert averages less than 100 mm (4 in) a year but can vary from 0 to 500 mm (20 in). Winters are cool but the summer heat is insufferable, climbing as high as 54°C (129°F).

Syrian Desert

Nation:	Syria
Expanse:	518,000 sq km (200,000 sq mi)
Average annual temperature:	from 10°C–46°C (50°F–115°F)
Rainfall:	130 mm (5 in)/year

Above: Ripples caused by prevailing winds that blow at the base of the dunes.

Right and on facing page: Extremely arid surfaces with scant vegetation.

The Syrian Desert occupies the northern part of the Arabian Peninsula, stretching northwest into Jordan, southwest into Syria, east into Iraq and south into Saudi Arabia. The desert cannot be considered strictly sandy, given its rock tablelands and vast pebble deposits. It is an unending and monotonous series of plateaus with variable heights from 300–600 m (984–1,969 ft). Many wadi drain toward the Euphrates River and cut deep canyons through a few areas.

The Syrian Desert is abnormally inhospitable. Rains are scarce and sporadic, and the quantity of water that precipitates or rises from underground does not permit the cultivation of any crops except in rare oases scattered around the region. Already sparse annual rains dwindle progressively in intensity as

one travels south, and they reach their minimum in a mountainous region called Al Hamad. Dry winds called khamsin blow twice a year in these areas as they do in the Egyptian Sahara. Carrying enormous sand loads, they are violent enough to raise walls of dust—the only particle small enough to be suspended at great heights— more than 1,000 m (3,028 ft) high that obscure the sky.

Nomad populations are concentrated in isolated oases where they raise camels and the renowned Arabian horse.

The most famous oasis is Palmyra, a name coined by its ancient Roman governors that means "city of palms." It is also known as Tadmur, a pre-Semitic word for "bride

of the desert," and is considered the heart of the Syrian Desert. It is located 210 km (130 mi) northwest of Damascus near hot-water springs that render the place a perfect rest stop for caravans traveling the routes that connect Iraq with Al-Shan—a name

for the assembled territories of Lebanon, Syria, the Holy Land and Jordan—over which silks were once transported from China to the Mediterranean. Palmyra offers con

siderable historic and artistic interest including architecture that represents the fusion of Greek and Roman cultures. A violent earthquake in 1089 devastated Palmyra and demonstrated the area's unsteady geodynamic structure.

Negev Desert

Nation:	Israel
Expanse:	12,000 sq km (4,630 sq mi)
Average annual temperature:	38°C (100°F)
Rainfall:	<100 mm (<4 in)/year

offers a great variety of lithologies and morphologies—sand dunes, varicolored sandstone deeply etched and sculptured by erosion, sheer canyons, wadi ready to fill

Right: Tafoni and rocks sculpted by meteorological agents near Timna.

Facing page: "King Solomon's Pillars" crafted by nature in sandstone rock.

The Negev Desert extends across Israel for 12,000 sq km (4,630 sq mi), or more than half the country's landmass. Its size is about one-sixth the expanse of the nearby Sinai Desert. Triangular in form, the Negev juts out toward the Mediterranean Sea and is bordered by the Dead and Red Seas from Beersheba to the port of Eilat.

Climatic conditions that are unmistakably arid, but especially stable, have permitted the continuous habitation of this desert for more than 60,000 years, longer than in any other desert in the world. Notwithstanding its "limited" size, the Negev

with water from one moment to the next, rocky expanses of rounded pebbles, and finally, singular morphologies that are only found here.

Plateaus in the center desert are true flat mountains that rise for more than 1,000 m (3,280 ft). A succession of raised zones and depressed areas eroded by hydrological and meteorological action have formed five wide semicircular valleys or oval craters similar to those created by the impact of meteorites. These "erosion valleys," called *makhteshim*, are desolate landscapes that conceal subsurface water. Makhtesh Ramon is the most spectacular from a geomorphologic standpoint. One interesting feature of Makhtesh Ramon consists of swarms of intrusive volcanic dikes that look like vertical rock layers inside pre-existing rocks that are layered horizontally. A structure called "Carpenter Hill," a promontory of columnar basalt several hundred meters high is also distinctive. The wadi and layers of sandstone colored green to brown and yellow to red are likewise spectacular. The town of Mitzpe Ramon is about 900 m (2,953 ft) above this depression. It is home to a geological museum whose headquarters rise directly over the Makhtesh Ramon crater itself. The Eilat region with its namesake mountains lies in the southern part of the desert near the Red Sea. The land here has been disrupted by tectonic events linked to the formation of the Siro-African fracture, which continues along the continent and gives rise to the great African Rift Valley. The lithology exposed by erosion is kaleidoscopic, from red granite of intrusive origin to yellow limestone and on to sandstone. Natural arches, mushroom rocks, tafoni and canyons are on display. The region is rich in copper deposits; it has been mined since antiquity and most famously at King Sol-

omon's Mines, which is now a national park.

The northern part of the desert borders the small Judean Desert. Its mountains descend toward the Dead Sea from altitudes over 1,000 m (3,280 ft) and extend 400 m (1,310 ft) below sea level, the world's lowest elevation. Just before reaching the seashore, the desert takes one last leap over a 200-m (650-ft) high threshold that runs

for more than 6,000 km (3,730 mi) from Turkey to Ethiopia. Because mountains protect it, this part of the desert is one of the driest in the whole region. Any humidity in the territory comes directly from mountain rains that reach arid lands by way of the wadi. The desert transitions into the fertile valleys of Israel's coastal plain a little north of Beersheva, the capital of the Negev.

The Negev's flora and fauna are particularly rich. Its flowers, including several varieties of iris, bloom after rare rains, especially in March and April. Its animal life includes leopards, fennec, gazelle, oryx (gemsbok), ibex (Nubian goat) and a distinct, endemic species of zebra. All of them live in protected areas where they are raised according to a national program that pro-

vides for their reintroduction into native environments. An abundant variety of birds passes through during the migratory period, including pelicans, storks, gru, cormorants, and even a few pair of eagles and buzzards.

Central Asia hosts a desert region that, even though discontinuous, is one of the largest on Earth. Its arid conditions arise from its unusually large distance from the ocean's coasts and humid winds.

China's deserts cover about 1.1 million sq km (424,710 sq mi) and occupy about 11.5% of the country's total landmass. They lie in the Temperate Zone between 75° and 125° longitudes East, and between 35° and 50° latitudes North. Typical of the extremely arid conditions found within this vast area are those of the Taklimakan Desert in the Tarim basin. Asian deserts are often divided, according to local terminology, into "gobi" if rocky or gritty, and "shamo" if sandy. The Turkestan Desert is located between 36° and 48° latitudes North and 50° and 83° longitudes East. It borders the Caspian Sea on the west, the mountains that fringe Iran and Afghanistan on the south, the mountains adjacent to Sinkiang on the east, and the Kyrgyzstan Steppes on the north. The two great Karakum and Kyzylkum ergs lie in its interior.

The deserts of China began to form in the Upper Cretaceous when a subtropical high-pressure cell influenced the area. Later, with the uplift of the Tibetan plateau and the Himalayan orogeny in the Tertiary, the climate's continentality was accentuated, the monsoon system stabilized, and northern China became even more arid. The lakes in the Tarim and other basins gradually dried up and disappeared while the Taklimakan and other sandy deserts broadened considerably. The region's aridity was further accentuated by continued uplift during the Pleistocene and Holocene.

One of the most famous features of the Chinese deserts and all of central Asia is the presence of thick and widespread deposits of loess, or dust transported by wind then deposited into thick accumulations. The dust comes from the Gobi and Ordos Deserts. Soil cores pulled from the loess (up to 335 m/ 1,100 ft deep) and dated with various methods like paleomagnetism help us reconstruct the climatic history not only of this region but also for the entire globe. The oldest loess in the Central Loess Plateau is about 2.4–2.6 million years old and documents the drastic climatic change that coincides with the beginning of the Ice Age.

In cores taken from the Luochuan area, 17 periods of loess deposition have been identified that correspond to glaciation-related climate changes in the last 1.67 million years, though this does not suggest an absolute correlation between a glaciation and the deposition of a loess layer. There may have been a greater frequency of cyclonic depressions and sandstorms in the Gobi Desert during a glacial period in the mountains of Tibet, Tian Shan and Kunlun. This would have induced a more effective transport of dust toward the east by the western jet stream, which was confined north of the Tibetan anticyclone. The accumulation of loess ceased throughout central Asia shortly before the Holocene.

ASIA

KAZAKHSTAN

MONGOLIA

CHINA

INDIA

Tropic
of Cancer

37. Karakum and Kyzylkum Deserts

36. Gobi Desert

31. Tengger Desert
32. Ordos Desert
33. Turpan Desert
34. Taklimakan Desert

Equator

35. Thar Desert

Indian Ocean

The Aral-Caspian basin in the western Turkestan Desert shows evidence that it was occupied during the ice ages by the largest known pluvial lake, which covered an area of 1.1 million sq km (424,710 sq mi).

The Great Indian or Thar Desert is located in northwest India, extending from the Arawalli Range east of the plain of the Indus to the mountains of Beluchistan on the west. The Thar's age and origin are still debated, and according to some, it is only as old as the Holocene, or younger than 10,000 years; others believe it began to form during the Middle Pleistocene. The desert is spanned by the Indus and its tributaries and crossed by fossil riverbeds. New evidence indicates that these fossil riverbeds are not always linked to ancient fluvial flows but may, in fact, be caused by tectonic activity. NASA's Earth-observing Terra satellite passed over the desert before (January 15) and after (January 31) the powerful earthquake tremor that struck on January 26, 2001 with an epicenter in the southern Thar. The shaking of the terrain caused groundwater to surface, forming a large basin and numerous drainage channels.

Tengger Desert

Nation:	China
Expanse:	42,700 sq km (16,486 sq mi)
Average annual temperature:	−2°C–16°C (28°F–61°F)
Rainfall:	250–500 mm (10–20 in)/year

This satellite image shows snow on the Kunlun and Tien mountains (along the top of the photo), whose glacial waters dislodge material that forms broad cones of alluvial detritus plunging all the way to the desert. A dust storm is visible at the foot of the mountains, on the left.

The Tengger is located on the Alashan Plain in north-central China. It terminates against the Qilian Shan Mountains in the south, while the Huang He (Yellow River) borders it on the east; it blends into the Gobi Desert on the north. Prevailing winds blow from the northwest quadrant and determine the migration of dunes toward the Huang He. Only 7% of the dunes created in this huge area can be considered stable. Village inhabitants populating the edges of the region exploit a playa with salt deposits at the center of the desert.

At the edge of the Tengger, the Sino Academy built a scientific station (Shapotou Desert Research Station) to monitor the dunes' migration and study control methods that will limit damage to the trans-desert railway. Rows of vegetation have been planted on a grid whose mesh measures a little over 1 sq m (3 sq ft). This system has made it possible to avoid repeated interruptions of the rail line in an area where the dunes move southeast by as much as 15 m (50 ft) a year. The vegetation has reduced the wind's ground speed by 17%.

The dunes of the Tengger Desert, like those in other Chinese deserts, can be traced to the Upper Cretaceous or Lower Tertiary Periods (about 65 million years ago). During that time, the whole area experienced subtropical conditions under a band of high pressure analogous to current conditions in the Sahara. Later in the Tertiary, when

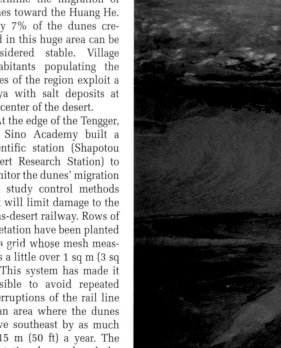

the Himalayan orogeny further lifted the Tibetan plateau, this climatic pattern was strongly accentuated. The monsoon system developed, and northwestern China dried out even more. The ancient lakes in various basins dried out gradually and were replaced by sandy deserts. The mountains' continued uplift during the Pleistocene and Holocene further accentuated the area's aridity. Proof of climatic fluctuations during the Pleistocene and Quaternary are preserved in these desert basins in loess deposits, as in other central Asian deserts.

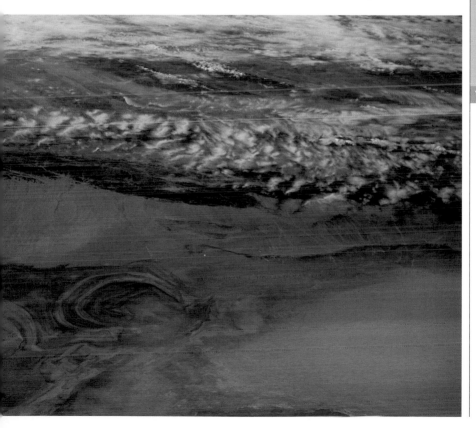

Ordos Desert

Nation:	China
Expanse:	32,000 sq km (12,355 sq mi)
Average annual temperature:	0°C–20°C (32°F–68°F)
Rainfall:	>250 mm (>10 in)/year

This satellite image focuses on territory occupied by an arid steppe. The course of the Huang He (or Yellow) River crosses the desert where vast accumulations of sand are deposited. White marks on the photo show the presence of salt lakes.

The Ordos is a "mountain" desert separated from the Tengger by the Huang He (Yellow River), reaching a maximum elevation of 2,000 m (6,562 ft). Its dunes are typically low (between 3–5 m/ 10–16 ft high) and the landscape in general seems only lightly modified by fluvial erosion. Given the presence of archeological remains and other evidence of ancient climates more humid than the current one, Chinese scientists propose that some deserts in China's central and eastern regions must be located strictly in relation to the presence or direct action of man. In this context, the Ordos Desert is frequently cited as one of the most desertified areas in China. Communities of herders have worked the alluvial plains along the Huang He since the ninth century B.C. Their poor agricultural practices destroyed the delicate balance of the steppe and allowed runaway desert expansion. The mobile dunes of the Ordos now cover the ruins of the capital of the Xia Dynasty (2205–1766 B.C.) and another 11 large cities.

Today, vast sandy areas and many small lakes, some of which are saline, characterize half of the Ordos region. Swamps and agricultural lands make up the other half. Roused by a will to reclaim their land, inhabitants of the Ordos have developed a special method for stabilizing the dunes. They have planted bushes on the lowest third of the dunes'

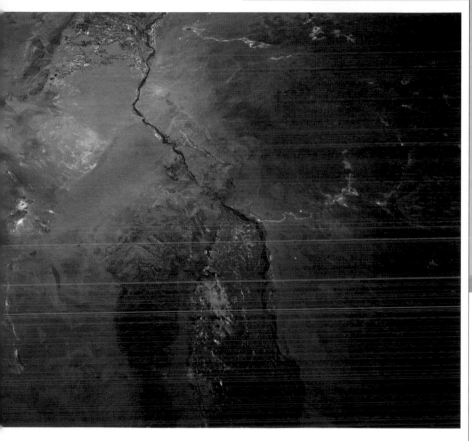

windward flanks. This vegetation lowers wind speeds at the base of the dunes and keeps most of the sand from moving toward the crest of the dune. A high wind velocity flattens the crests that are no longer fed with sand. At this point trees can be planted on the flattened surface of the dunes. In five years, this method can increase the vegetative cover by 50–80%. Another technique used in many sandy areas of the Ordos (and other regions of China) to reclaim the desert is the construction of kulun. These are fences that allow grazing to be controlled and the land to recover from the abuse it has suffered. Pastureland and forests slow the wind and improve the utility of the land inside the kulun. The kulun can be built to provide forage, control the movement of sand, or to convert marshy areas among the dunes into lands suitable for agriculture.

Turpan Desert

Nation:	China
Expanse:	50,000 sq km (19,305 sq mi)
Average annual temperature:	−10°C–32°C (14°F–90°F)
Rainfall:	30 mm (1.2 in)/year

This satellite image shows the lofty Tian Shan Mountains at the northern border of the Turpan Desert. Clearly visible above the fluvial deposition cones at the base of the mountains is a fierce sandstorm with winds of 100 km (60 mi) per hour.

The small Turpan Desert lies in the self-governing Xinjiang region of northwest China. It ranges south of the Tian Shan Mountains and is bordered by the Po-ko-ta chain on the north and the K'u-lu-k'o-t'a-ko Mountains on the south. The Turpan lies 150 m (490 ft) below sea level, the lowest area in China, and is known, appropriately, as the Turpan Depression. The Chüeh-lo-t'a-ko Mountains divide the desert basin in two parts. The northern section is situated at the feet of the Po-ko-ta and drains south through deep, branching canyons into what was once a permanent lake but is now completely dried out and home to salt lakes. These small basins are only occasionally inundated; the most famous and representative among them are the Ai-ting and the Aydingkol Hu. The latter, whose floor is situated 154 m (505 ft) below sea level, represents the second lowest elevation on Earth. Northern winds are largely blocked by the stately Tian Shan Mountains, but manage to escape through a few passages and sweep through the Turpan Depression. These areas are then more exposed to powerful erosion than to deposition and are largely composed of quartzite pebbles. Most of the sand is exported from this locality, being transported for long distances and deposited in a 2,500-sq-km (965-sq-mi) area within the depression called the "Mountain of Sand." This structure's morphology and the distribution of its dunes are

dominated by the prevailing northeast winds, while the weaker northern and north-west winds cannot penetrate this far. The dunes are linear and disposed along fairly regular lines separated by sand flats. Stellar dunes predominate in the area's north-eastern region, which proves that winds of equivalent strength are blowing in from different directions. Dunes characterize the interior zone,

and its powerful sandstorms force local inhabitants to defend their buildings with barriers that reduce the force of the wind. Fifty years ago, a storm with wind velocities over 130 km/h (80 m/h) carried enough sand to bury every house.

Extreme temperatures characterize the Turpan. The monthly average is –10°C (14°F) in January and 32°C (90°F) in July with consistent

daily and monthly excursions; the highest temperature ever recorded was 48.9°C (120°F), while the lowest of –52°C (-61°F) was recorded in Fuyun, not far from Turpan. Locked in a natural basin and surrounded by enormous mountains, the area is well protected from weather disturbances and is scarcely affected by rains. Annual precipitation here is only 16–30 mm (0.6 to 1.2 in). Abundant water flows from the high mountains of the Tian Shan chain and is withdrawn directly by the inhabitants, who intercept and capture it at the higher altitudes by digging tunnels that carry it into the basin.

Taklimakan Desert

Nation:	China
Expanse:	323,750 sq km (125,000 sq mi)
Average annual temperature:	0°C–18°C (32°F–64°F)
Rainfall:	<20 mm (<0.75 in)/year

The Taklimakan Desert occupies the basin of the Tarim River (*Tarim Darja* or *T'a-li-mu Ho,* in Chinese). The Tien Shan Mountains rise directly north of the alluvial plain,

The Kunlun Shan Mountains mark the southern border of this desert.

while the southern mountains of the Kunlun Shan and the eastern Karakorum chain complete its perimeter. Toward the east and northwest, the Taklimakan merges with the Tengger and Gobi deserts. The Tarim River originates in the Tien Shan Mountains and flows west, crossing the northern part of the basin. Its course terminates in a depression in the closed drainage basin of Lop Nur, but its waters usually evaporate long before they reach the ephemeral lake in the middle of the depression. The

Qarqan River (Ch'e-erh-ch'en Ho, in Chinese) drains the mountains bordering the basin on the south and, like the Tarim, its waters evaporate before they reach Lop Nur.

Barchans in the Takli-makan Desert are among the largest in the world. The sands of the barchans origi-nate in the Tarim's alluvial plain and the mobile barchans migrate in a westerly direc-tion. The distance between a barchan's two horns averages 3.2 km (2 mi), and its total average length is 2.2 km (1.4 mi). The distance from one barchan to another is 3 km (1.9 mi), and they average 100–150 m (330–490 ft) in height.

Playas and interdune lakes coincide with small chan-nels of the Tarim. When geo-morphologists studied this area in 1905 and counted 35 lakes, the locals reported that these were created artifi-cially by taking advantage of interdune depressions. The researchers further ques-tioned the true origin of these lakes, however, and found that some of them were fed by a system of channels from the Tarim, so they concluded that they once contained fresh water, which turned brackish when the channels dried up. They also noted that the lakes had to be young since the Tarim only recently developed frequent course changes. The lakes' duration depends upon both the river's instability and the movement of the dunes. The

Below: A section of the River Tarim.

Facing page: The last outposts of the saline Lop Nur lake basin.

river's instability was also documented when satellite radar images revealed channels of the Tarim beneath the sand, 80 km (50 mi) south of the river's current channel.

The Lop Nur lies in the eastern part of its basin at an altitude of about 780 m (2,560 ft). It is a wide, salt-encrusted playa into which the Tarim drains. The Lop Nur is reported on ancient Chinese maps and in other documents as a salt lake more than 150 km (90 mi) wide. Currently, it is completely dry and surfaced with a salt crust 30–100 cm (12–40 in) thick. Concentric rings created by the lake's progressive withdrawal look like a gigantic bull's-eye when viewed from space. Average annual precipitation is below 20 mm (0.8 in) but water will sometimes stagnate for short periods in its center.

Yardangs were first described in this desert; they are peculiar elongated landforms sculpted by the wind (and already described in this volume on page 64). Calculations suggest that more than 5 m (16 ft) may have eroded from their surfaces in the Taklimakan between 1919 and 1959, even though this was probably a local phenomenon. In addition, yardangs developed in the direction of the Lop Nur's ancient high waters, and some evidence of water erosion can be observed. If this is true, it is possible that not only wind but also water played a sig-

nificant role in the genesis of these landforms.

The ruins of Kroraina (also known as Loulan) lie a little northwest of Lop Nur. Kroraina was an important stop on the Silk Road from the fourth century A.D. when fish from the Lop Nur's tributaries, domestic herds and wildlife from the surrounding countryside fed its more than 10,000 residents. Today, the area is a desolate plain that even now can only be

reached by camel. Few ruins are left to mark the ancient and flourishing metropolis.

Thar Desert

Nation:	Western India–Pakistan
Expanse:	214,000 sq km (82,625 sq mi)
Average annual temperature:	–10°C–32°C (14°F–90°F) July
Rainfall:	100 mm–400 mm (4 to 16 in)/year

Above and right: Great expanses of sand stretch hundreds of kilometers from the slopes of the Arawalli Mountains.

The Thar Desert is an expanse of sand 214,000 sq km (82,625 sq mi) in size that overlays the ancient alluvial plain of the Indus River. The Arawalli Mountains confines it on the east, while it borders the current plain of the Indus itself on the west. Its Indian name is Marwar, meaning "place of death," but it is not extremely arid since precipitation drops below 100 mm (4 in) a year only locally.

The Thar is a monsoon desert. During the summer, the heating of the northern part of the Indian subcontinent controls a deep low-pressure weather pattern, which is anchored in place by the Himalayas. Under these conditions, the pressure gradient causes the air and humidity to move toward the low. In September, the winds change direction and the northeast monsoon begins to blow. Winter winds come from the cold continental anticyclone. Wind speed reaches its height in May and June (corresponding to minimum temperatures), while rain is at its maximum in July and August.

An important feature of the Thar is the presence of

Himalayan glacial- and snow-melt waters, transported now as in the past by the Indus and its tributaries. Seasonal flooding can be truly destructive. The annual discharge of the Indus is almost two times that of the Nile, and it can reach 16 km (10 mi) in width at high water. Sands and clays transported by the river over the last 5,000 years are about 10 m (33 ft) deep on the alluvial plain. Numerous channels that represent ancient secondary watercourses cross the Thar.

Below and right: Great dunes march endlessly across the Thar Desert, which collects the greatest quantity of sand in the world.

The Thar is noteworthy for the nature of its dunes. Enormous amounts of sand of fluvial origin are the dunes' source material. Clusters of parabolic dunes with multiple arms are unique to the region. They reach from 10–70 m (33–230 ft) high and 2.5 km (1.5 mi) long and wide on average. These are particularly well developed north-east of Naya Chor in the middle of the basin. As a whole, these dune clusters make up a rake-shaped structure. They grow where vegetation anchors the arms and are only minimally bound to the force of the wind because

the Thar is one of the least windy deserts on Earth.

Linear dunes as high as 30 m (100 ft) develop west of the areas with parabolic dunes. These linear dunes are complex structures since secondary bodies develop on their crests and windward flanks, which diverge from a convex center that develops perpendicularly to the southwest summer monsoon. It is, therefore, possible that the linear dunes develop as a consequence of wind eroding the parabolic dunes' inner body. Starting from an inactive channel of the Indus called the Hakra, linear dunes

develop on their leeward flanks. Interdune lakes develop between the linear dunes and, in some cases, at the boundary between some of these and the complex dunes. Salt deposits are observed in some locations and there are sabkhas near the coast.

Dunes began to develop in the Thar during the last glacial period when monsoon circulation was weaker. They stabilized when the climate began to heat up again during the Holocene and the monsoons regained their strength. The dunes may have been covered by vegetation during

the first phases of their growth. Underground springs in the southwestern region have given rise to agricultural practices that take advantage of these waters.

Gobi Desert

Nation:	Mongolia/China
Expanse:	1,300,000 sq km (501,930 sq mi)
Average annual temperature:	−18°C–20°C (−0.4°F–68°F)
Rainfall:	76 mm (3 in)/year (western regions)
	203 mm (8 in)/year (northeastern regions)

Above: A skeleton of a Tarbosaurus, a carnivorous dinosaur found in the desert's sandy zone.

Right: A zone of transition to the most desert-like area where no vegetation takes root.

The Gobi Desert covers a vast region of central Asia that extends through Mongolia and the Mongolian interior within the Chinese Republic. It occupies an east-west arc about 1,600 km (990 mi) long and 480–965 km (2985–600 mi) wide. Its approximate boundaries are the Altaj and Hangaj Mountains on the north, the Great Khingan chain on the east, the A-erh-chin, Pei and Yin Mountains on the south and finally, the eastern Tien Shan on the west.

The Gobi is composed of vast limestone plains that date to the Cenozoic era about 65 million years ago. The monot-ony of the landscape is interrupted by hills and rocky crests that date from a more ancient time. The terrain is composed of bare limestone that is covered in places by mobile sand. Gobi soils, where present, are gray-brown and made of chalk, large grit and sand. These immense rock

plains are essentially free of streams, but there are small lakes and salt bogs in its depressed areas.

An interesting museum in Dalari-Dzadgad, the main city on the Gobi plateau about a two-hour flight from Ulan Bator, offers abundant documentation of the first paleon-tological studies undertaken in the Gobi. The research showed the Gobi to contain an especially rich collection of dinosaur remains. Many forests, lake basins and streams characterized the Gobi in the time of the dinosaurs, 65 to 130 million years ago. Today, the only visible evidence of the ancient forests is a large fossil trunk outside the museum.

The Gobi is subdivided into several regions separated by mountain zones: the basins of the Ka-shun, Trans-Altaj and Dzungaria to the west; the eastern Gobi or Mongol in the center-east; and the

Ala Shan Desert in the south. The Ka-shun, bordered by the Tian Shan and Pei Shan mountains, rises to an altitude of 1,500 m (4,920 ft) and has an irregular and complex morphology, with wide depressions separated by flat hills and rock crests up to 90 m (300 ft) high. The Trans-Altaj, located between the Mongol and Gobi Altaj, is a harsh, elevated plain with

Above: Extreme changes in temperature cause serious damage to vegetation.

Right: A Tarbosaurus found in the desert and exhibited at the headquarters of the Academy of Sciences in Ulan Bator.

Center: The last forests on the edge of the desert.

isolated groups of low, rounded hills and mesas and crossed by dry riverbeds and steep slopes. The Dzungaria lies north of the Tien Shan chain and is a large basin with a southwest inclination at 50–70 m (160–230 ft) below sea level. It collects water from streams that descend from the Altaj in the north and the Tien Shan in the south that form a large num-

that slopes to the northwest and is largely covered in sand.

The elevation of the eastern Gobi varies from 700–1,500 m (2,300–4,920 ft). It receives more rain (up to 200 mm/8 in a year) than the other Asian deserts, but it has no other rivers than the Kerulen. The climate is dry and continental, meaning hot summers, cold dry springs and frigid winters. The tem-

The Gobi's extreme aridity depends on two factors. The terrain is primarily limestone and thus capable of capturing and swallowing the rivers that run across it, which are rare in any case, and the imposing mountain structures found to the west furnish protection from rain. Even in the most arid regions there is always water in the subsoil, and where the rains

ber of briny marshes in its interior. The Ala Shan is situated between the Chinese-Mongolian border on the north and the Yellow River on the south; it is a vast plain

perature varies considerably during the year with January lows of –40°C (–40°F) and highs of 45°C (113°F) in July. Daily thermal excursions can also be significant.

are more conspicuous, the relatively abundant subterranean and surface waters feed small lakes and springs.

Karakum and Kyzylkum Deserts

Nation:	Turkmenistan/Uzbekistan/Kazakhstan
Expanse:	Karakum: 490,000 sq km (189,190 sq mi)
	Kyzylkum: 200,000 sq km (77,220 sq mi)
Average annual temperature:	−30°C–54°C (−22°F–129°F)
Rainfall:	<150 mm (<6 in)/year

A satellite image shot in June 1985 shows that the Aral Sea then covered about 67,000 sq km (25,870 sq mi). The sea is currently shrinking due to the diversion of water from the Amudarya, one of its major feeder rivers.

The Karakum and Kyzylkum deserts comprise one-quarter of Earth's desert surface. Their combined area is 690,000 sq km (266,400 sq mi), and they are 1,280 km (796 mi) long east to west, and 960 km (597 mi) north-south. Their eastern border is the hilly piedmont landscape of the Pamir and Tien Shan mountains. Two long rivers, the Amudarya and Syrdarya, descend northwest from the glaciers of these mountains and empty into the Aral Sea.

The Amudarya River, described as the *Oxus* by Herodotus in the fifth century, A.D., is the longest river in central Asia at 2,550 km (1,586 mi) long. When Herodotus wrote about it, the river ran toward the Caspian Sea and only later changed course toward the Aral Sea. The Amudarya at first flows toward the Karakum's central depression, known as the Sarykamysh Depression, before emptying into the 64,000-sq-km (24,710 sq-mi) lake via its large delta. The region has an average rainfall of less than 1,100 mm (43 in), of which about 1 m (3 ft) evaporates each year as lake water and is replaced by water from the lake's two tributary rivers. Withdrawal of irrigation water from the two rivers in recent decades has lowered the level of the Aral Sea and its waters are becoming salty.

The Amudarya's average flow is 1,500 m³/s (53,000 ft³/s), but during summer high waters in June and July and coincident with maximum glacial runoff from the mountain, it reaches 3,380 m³/s (120,000 ft³/s). The flow of solids exceeds 100 million tons as a suspension load and 5 million tons as bed load. All this material is dropped on the delta and adds to the sedimentation of the Aral Sea. The eastern part of the delta has been active since the tenth century, while the current course of the river stabilized between the end of the 17th century and the beginning of the 18th. Much of the delta plain is occupied by salt playas.

During the last Ice Age, the largest known pluvial lake occupied the area between the Caspian and Aral seas, along with much of the Karakum. The Caspian and the Aral were then united in a single lake that covered 1.1 million sq km (425,000 sq mi).

The shoreline was higher than that of today's Caspian Sea by 76 m (250 ft), and the lake stretched 1,300 km (808 mi) north into the Volga River

lies northeast of the Caspian between it and the Aral Sea and is a desert with no dunes and made solely of clay eroded by the wind and ancient streams. It is also completely free of vegetation except in rare oases. The Barsa Kel'mes Depression is completely cloaked in salt and a place of no return southwest of the Aral Sea.

The Kyzylkum, or red (*kyryl*) sands (*kum*), is made of plateaus and hills formed by sand mixed with grit. There are also extended fields of linear dunes.

Climatic conditions of the two deserts are extreme. Summer temperatures can reach 54°C (129°F) while winter temperatures oscillate between −25°C (−13°F) and −30°C (−22°F) when the deserts are exposed to strong cold winds from Siberia. Annual rains average less than 6 inches per year, but in some cases barely exceed 10 mm (0.4 in) a year. Sandstorms occur an average of 60 days a year and are especially violent and persistent.

plain. Deserts took shape from the desiccation of this lake. The Amudarya divides the former pluvial lake basin into two deserts, with the Karakum on the west and the Kyzylkum on the east. For the local populations, the Amudarya is the mother river and the two deserts are her two children.

The Karakum, or black (*kara*) sands (*kum*), is composed of a denuded flatland with abundant salt plains, erosion furrows in clay deserts (called *takir*, some of which cover hundreds of square kilometers), and areas with karst topography. The sands of the Karakum come from ancient lake bottoms and alluvial plains that were formed during the last several episodes of glaciation. Barchans cover approximately 6.8 million hectares (16.8 million acres), or 10% of the Karakum. Winds from the Caspian Sea transport salt and deposit loess. Due to continuing tectonic activity near the Caspian at the edge of the desert proper, there are thermal springs and mud craters that emit water rich in metal oxides. The Usjurt plain and the Barsa Kel'mes Depression lie on the margin of the Karakum. The Usjurt

AMERICAS

The deserts of North America occupy a large part of the western United States and northern Mexico between 22° and 44° latitudes North. They extend south from Oregon, including almost all of Nevada and Utah as far as southwestern Wyoming and western Colorado, and advance west into southern California up to the eastern base of the Sierra Nevada, the San Bernardino Mountains and the mountains of Cuyamanca. From southern Utah, the deserts continue into Arizona and the Chihuahua Desert in Mexico. California's Sonora Desert extends into Baja California and along the eastern side of the Gulf of California. Cumulatively, these regions constitute a single, large, if somewhat discontinuous, desert.

The regions comprising the North American deserts owe their aridity to different causes. The barrier offered by the mountains places a certain importance on the more northern areas and on California, while the southern part of the deserts is subject to the influence of a subtropical high-pressure cell and sees its highest precipitation in the summer. The extreme aridity has a very limited scope; the most arid regions are situated along the Gulf of California and in the Mojave. Fluvial activity in the hyperarid zones is intense, however, due to streams that flow from the nearby mountain chains.

The genesis and climatic evolution of North American deserts depend largely on the orogenesis of the cordillera that started during the Cretaceous and is still active. The uplift led to the rivers' cutting deep and spectacular canyons.

The Basin and Range Province, which hosts much of these deserts, began to develop in the Upper Oligocene due to a crustal extension. It is composed of an ensemble of raised masses, bound by faults and closed, elongated depressions. These depressions host playas, one of the most typical features of North American deserts. These are the remnants of vast pluvial lakes that formed on several occasions under different climatic conditions. During the last Ice Age, more than 100 closed depressions accommodated lakes but only 10% remain today. Lake Bonneville was the largest of these lakes at 500 km (310 mi) long, 335 m (1,100 ft) deep, and covering a maximum of 51,640 sq km (19,940 sq mi). Lake Lahontan was second largest, covering a surface of 22,900 sq km (8,840 sq mi) to a depth of 228 m (748 ft).

The history of these lakes is well documented by sediment profiles taken from the bottom of the playas, yet the causes of such drastic climatic and environmental changes are not known with certainty. It may be the case that reduced temperatures caused evaporation to diminish or that rainfall increased. In addition, there is geologic evidence documenting that the deserts once covered far more land than they do now. Large systems of paleodunes have been described east of the Western Cordillera between the Canadian border and the Gulf of Mexico, in Nebraska, and on the

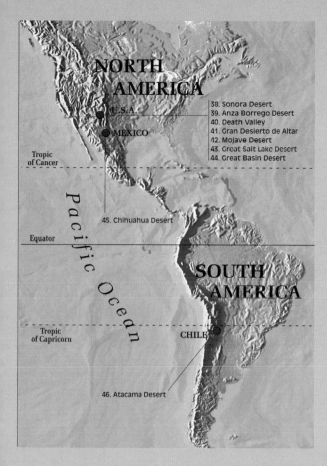

NORTH AMERICA

U.S.A.

MEXICO

38. Sonora Desert
39. Anza Borrego Desert
40. Death Valley
41. Gran Desierto de Altar
42. Mojave Desert
43. Great Salt Lake Desert
44. Great Basin Desert

Tropic of Cancer

Pacific Ocean

45. Chihuahua Desert

Equator

SOUTH AMERICA

Tropic of Capricorn

CHILE

46. Atacama Desert

---

high plains of Texas where the desertification phase is 1.4 million years old.

The primary desert areas in South America are linked to the Andes Cordillera. The largest zone includes Peru's coastal desert west of the mountains between 5° and about 30° latitudes South. The Argentine deserts of Monte and Patagonia lie east of the Cordillera. Both Argentine deserts are located on the Andes' leeward flank. The Monte Desert, in essence a continuation of the western deserts, has a "basin and range" topography with many closed depressions. The Patagonian Desert extends for more than 500 km (310 mi) between the Andes and the ocean and owes its aridity to the cold Falklands current offshore and the mountains that block humid western winds. Pediment plains that slope east toward the Atlantic dominate the region. During the Plio-Quaternary, the Patagonian Desert suffered climatic modifications significant enough that the advancing Andean glaciers deposited the widespread glacial and fluvial sediments on its surface.

The desert between Peru and Chile is different. Climatically, its aridity is generated by the tropical atmospheric high pressure reinforced by the upwelling of cold coastal waters associated with the north-flowing Peruvian Current. This is one of the most arid zones on Earth even though the coastal zone is characterized by fog (*camanchaca*) from the Pacific. The landscape is dominated by pediment surfaces.

The evolution of these deserts is still a controversial issue. The uplift of the Andes was mild enough following the Upper Miocene (about 10 million years ago) that its rivers have only deepened their canyons by 100–200 m (330 to 660 ft). The Atacama Desert was already arid from the Upper Eocene onward (about 40 million years ago), but it became hyperarid only with the raising of the Andes. As with Africa's Namib Desert, this occurred sometime between the Oligocene and Miocene and followed the evolution of deep cold waters from Antarctica and the cold Humboldt Current between 13 and 15 million years ago.

Sonora Desert

Nation:	United States/Mexico
Expanse:	310,000 sq km (120,000 sq mi)
Average annual temperature:	–9°C–40°C (16°F–104°F)
Rainfall:	0–130 mm (0–5 in)/year

On these pages: Extreme aridity characterizes the Sonora's desert landscapes; polygonal mud surfaces, the work of desiccation, are plainly evident.

(120,000 sq mi) in southwest Arizona, southeast California, much of Baja California and in the western swath of the Mexican state of Sonora. Many would include the Sonoran within the Colorado Desert (or *Colorado Sonoran*), the Yuma Desert and the Gran Desierto. Some suggest that it also includes the Mojave Desert (*Mojave Sono-*

Deserts beyond North America generally evolved on structural surfaces with horizontal strata or on ancient alluvial plains that stretched hundreds of kilometers. Many contain closed depressions with well-preserved lacustrine terraces that represent the remains of ancient pluvial lakes. Deep canyons like the Grand Canyon and mountain chains are associated with these deserts. Dunes are rare in North American deserts, and fluvial processes dominate their morphology.

The Sonora Desert is in a subtropical climate and is the most complex of North America's deserts. It is an arid region that covers 310,000 sq km

ran), which can only be separated from the Colorado Desert, however, on the basis of its higher altitude. On the basis of vegetation, the Sonoran can be subdivided into seven different deserts—or eight, if one considers the Mojave as part of the Sonoran. Two peak rainy periods annually, a short cold season, and geologic and topo-graphic diversity give the Sonora Desert a greater variety of plant life than is found in any other desert in the world. The western Sonoran or Colorado Desert and the nearby Anza Borrego Desert are well sheltered from Pacific storms and famous for the magnificent flowering of ephemeral plants following late-winter rains.

The Arizona Uplands is the area most unlike the rest of this desert. In fact, abundant humidity and elevation differentials as great as 600 m (1,970 ft) give rise to vast, gently sloping alluvial cones that accumulate at the base of mountains to form a continuous debris apron called a bajada.

The complex and varied

Above and right: Cacti are fat, luxuriant plants well adapted to the humidity and poor climate. The abundant saguaros represent most of the territory's vegetation.

and August. The summer "monsoons" are characterized by intense downpours during powerful storms that interrupt the desert's intense heat and give rise to an outburst of flowering and vegetative growth.

The Sonora Desert is bordered by mountains in Colorado, Arizona and Sonora that receive large amounts of precipitation in the form of rain and snow. Water streams from the mountains into rivers that cross the desert. Some of these rivers also support a corridor of vegetation during the drier seasons. The Colorado and San Pedro Rivers are perennial and flow throughout the year. The Gila River has been less constant since dikes were built along its course, and the Santa Cruz River is intermittent. In the upper Santa Cruz basin, however, rains feed subterranean waters sufficiently enough that the city of Tucson, Arizona, once relied exclusively on them for its municipal supply.

The plants of the Sonora Desert have perfected various adaptations to survive. For example, mesquite roots dig down as far as 30 m (100 ft) to reach water, while the roots of the creosote specialize in extracting almost all the water in the soil. The uniform spacing between creosote plants is the result of competition among their roots for the scarce water available in the subsoil. Ironwood (*Olneya*) produces leaves under favor-

vegetation of the Arizona uplands includes an abundance of succulent plants. The giant saguaro (*Carnegiea gigantea*) grows here. It is the symbol of the Sonoran and can grow to more than 15 m (50 ft) tall. Where conditions are favorable, saguaros grow like forests, interspersed with dense woods of mesquite (*Prosopis*), acacias and *palo verde* (*Cercidium*).

Precipitation in the Sonoran arrives in peak rainfalls during two different seasons—winter and summer—rather than one alone. Winter rains fall from December to March and arrive from the west, while summer rains approach from the Gulf of Mexico in the east in July

able humid conditions and loses them the moment those conditions change; photosynthesis is then activated in its trunk and branches. The ocotillo (*Fouquieria*) behaves the same way. To avoid competition, some desert shrubs like the brittlebush (*Encelia*) produce poisons that inhibit the growth of new plants around them. The seeds of the palo verde (*Cercidium*), ironwood (*Olneya*) and smoke tree (*Dalea*) are extremely hard and thus resist desiccation. The seedpods must be scraped away mechanically in order to germinate and that can happen in the streaming that follows violent downpours.

Some Sonoran plants have been imported from other areas. Among them are the salt-tolerant tamarisks (*Tamarix*) that were imported from the Mediterranean to control wind erosion. Parasites, insects and diseases that were not imported into the United States with the plants restrain tamarisk populations in their region of origin. For these reasons, the tamarisks, whose presence increases the salinity of the soil, have spread disproportionately and replaced local species. Programs for the control and eradication of this plant are being conducted throughout the southwestern U.S.A., but their complete elimination is not currently economical.

Anza Borrego Desert

Nation:	United States
Expanse:	2,400 sq km (930 sq mi)
Average annual temperature:	42°C (107.7°F)
Rainfall:	174 mm (7 in)/year

Carrizo Badlands, lose altitude and connect to the Salton Trough. The Borrego Valley and Borrego Badlands are among the hottest places in the United States. During

White roads cross the desert burned by the sun. A dust finer than sand is a typical feature of these areas.

The Anza Borrego Desert is located in southeastern California and covers 2,400 sq km (930 sq mi). It is bordered on the north by the Bucksnort and Santa Rosa Mountains, by the Jacumba Mountains on the south, the Pinyon Mountains on the west and the Borrego Mountains on the east, which then blend into the

summer their daytime temperatures average 42°C (108°F) and peak at 50°C (122°F) (June 1996). The average precipitation was 174 mm (7 in) a year between 1962–1994, but it was 0 mm during an episode of paramount dryness (1982-1989). Hurricane Kathleen brought about the maximum monthly precipitation of 128 mm (5 in) in September 1976.

The Anza Borrego Desert offers huge scenic variety, perhaps the richest in the world. This certainly is re-

peoples were semi-nomadic; they spent the winter in the desert lowlands and moved to higher elevations between late spring and autumn, following the harvest of acorns

Sea, the land slopes to 71 m (233 ft) below sea level. This depression took shape recently and is linked to California's especially concentrated tectonic activity. The best North

lated to its passage from areas below sea level to altitudes of 2,400 m (7,874 ft). Since 1933, the desert has been included in the Anza Borrego Desert (California) State Park. It contains abundant archeological evidence of populations that occupied this territory before the Spanish conquest. The ancient

and pine nuts. In some parts of this desert, and especially in the Santa Rosa Mountains, an ancient road network crosses the mesa. Stone piles called "cairns" along these trails have been interpreted as signposts that are analogous to the "*ometti*" used in the Alps. Going east toward the depression occupied by the Salton

American outcrops of Pliocene and Pleistocene sediments are visible in this area.

Death Valley

Nation:	United States
Expanse:	15,000 sq km (5,790 sq mi)
Average annual temperature:	40°C (104°F)
Rainfall:	<50 mm (<2 in)/year

Right: The desert depression known as Death Valley collects sand from the Amargosa and Panamint Mountains.

Facing page: The erosion-sculpted Sylvania Mountains limit Death Valley on the north.

Death Valley lies north of the Anza Borrego and Mojave Deserts in south-central California and Nevada. It was named by "49ers," the gold seekers who suffered great hardship attempting to cross it during the gold rush of 1849. Its ecosystem is analogous to the Mojave Desert's but geologically it is part of the Great Basin Desert. Death Valley is a vast desert depression encircled by mountains. It terminates at the Amargosa Range on the east, the Panamint Range on the west, the Owlshead Mountains on the south, and the Sylvania Mountains on the north. The surrounding mountains crest at 3,350-m (10,990-ft) Telescope Peak, and Badwater, the northern hemisphere's lowest elevation (86 m/283 ft below sea level) lies nearby, resulting in an elevation change in excess of 3 km (2 mi).

The climate of Death Valley is generally sunny, dry and clear all year long. Winters are mild with occasional disturbances, but the summers are torrid and dry. Some of the highest temperatures on earth have been recorded in Death Valley; they hit 56.6°C (133.8°F) on July 10, 1913, the highest temperature ever registered in the United States. It is completely normal for summer temperatures to surpass 48°C (118°F).

The Sierra Nevadas block most storms on the west, and are thus responsible for Death

Valley's extreme aridity. Average annual rainfall is less than 50 mm (2 in), and only 42.16 mm (1.64 in) fell in each of the last 50 years at Furnace Creek. High temperatures and low humidity effect an especially high rate of evaporation, which occurs at 3.25 mm (0.13 in) a year, or 77 times precipitation.

Death Valley is part of the Great Basin Desert and is similar to the other basins in the Basin and Range Province. It is unique, nonetheless, because at least 1,225 sq km (473 sq mi) of it are below sea level. Most of Death Valley's surface waters convene in salt ponds and swamps that surround the salt flats. For most of the year the Amargosa River bed is only a series of dry channels. The river carries a little bit of water in the southernmost part of the valley, but most of the water flow is subterranean. Salt Creek drains the northern part of the valley but only small stretches feature a perennial surface flow. During the Upper Pleistocene, a 180-m (590-ft)-deep lake occupied the valley and its evaporation created the current salt plain. The Cottonball Basin playa is a good example of a saline deposit that includes sodium borates and sulfates, chalk, rock salt, carbonates and lacustrine sediments. Some of these minerals were

once exploited, giving fame to the "twenty-mule team" freight wagon used to transport them.

Death Valley's geologic history is complex and involves tectonic and volcanic agents. It lies in a *graben* formed by the lowering of a crustal block along normal parallel faults accompanied by the uplift of *horsts* on the flanks of the Amargosa Range and the Panamint Range. The

linearity of the connection between the slopes of the Panamint and the Funeral Range with the floor of Death Valley underscores the existence of normal faults. The current structure and its topographic expression started to develop about 10 million years ago when the Furnace Creek fault system began to move. Miocene and Pliocene lake deposits indicate that lakes formed in the lowered

along the steep slopes produced by fault activity in the area. In many cases, the rock detritus has developed a dark coating of desert varnish. The fresher rocks laid down atop it are lighter in color, and the different tonalities make visible the distribution channels that have developed on many cones. The deepening of the depression led to conditions that favored its

On these pages: The masses of sand accumulate in high dunes that are sometimes colonized by the scarce vegetation.

zones. Contemporaneously, uplift, fault activity, erosion and volcanism produced massive amounts of detritus that accumulated in the depression that was under way. During the Pliocene (3 to 5 million years ago), tectonic activity became more intense and the current topography was produced.

Alluvial cones and bajadas were the result of erosion

replenishment by materials eroded from the surrounding mountains. The rock substrate in the center of Death Valley is buried under 2,750 m (9,000 ft) of sediment.

Despite its aridity and the severity of its environment, more than 1,000 plant species grow in Death Valley. Those living on the valley floor adapt to desert life in various ways. Some have roots that descend to a depth 10 times the height of a person; the root system of others grows just below the surface but in every direction; still others don spines that keep evaporation to a minimum. There are no plants (with some microscopic exceptions) on the salt flats, but plants like *Salicornia* that tolerate salt (halophiles) are widespread around the playas. Cacti are rare in the southern part of the valley, but more abundant on the northern bajadas.

Gran Desierto de Altar

Nation:	Mexico
Expanse:	310,000 sq km (119,690 sq mi)
Average annual temperature:	37°C (98.6°F)
Rainfall:	76–100 mm (3–4 in)/year

Facing page: A "mesa" or segment raised from the substrate and affected by intense tectonic activity linked to the San Andreas Fault.

The Gran Desierto is a coastal desert. More than half of its surface is less than 80 km (50 mi) from the ocean and it directly faces the Gulf of California. The Gulf of California was created about 12 million years ago by tectonic activity linked to the San Andreas Fault, and it cut the desert in two.

Tectonic activity left a characteristic imprint on the topography of this area similar to the basin-and-range topography of the Great Basin Desert. The delta of the Colorado River lies at the edge of the Gran Desierto and the upper end of the Gulf of California. In the Quaternary alone, the delta generated more than 7,700 sq km (2,970 sq mi) of new desert surface as it moved toward the Gulf. The Gran Desierto occupies a basin 60 m (200 ft) deep east of the Colorado River.

The mountain ranges lie in a northwest-southeast orientation both west of the alluvial-delta plain and east of the sand sea. The ranges feature sharp crests and huge alluvial cones at their base. Most Gran Desierto rock is igneous. The southern Sierra Pinacate, whose highest peak is the 1,000-m (3,280-ft) Cerro Pinacate, features a group of volcanoes that have been active for the last 2 million years (throughout the Quaternary) and are still active. The desert's sand sea occupies fields of volcanic lava that are alternately covered and uncovered by windblown sands. The Sierra del Rosario is surrounded by the sand sea in the middle of the desert, is about 25 km (15.5 mi) long and 5 km (3 mi) wide, and is made of granite with pegmatite dikes.

The dunes of the Gran Desierto that surround it average 180 m (590 ft) high, and their shapes vary from simple to compound and stellar. The complex star-shaped dunes found southwest of the Sierra del Rosario have three or four arms and succeed one other in parallel linear fascia that are clearly visible in satellite images of the area. To the east, this area of parallel dunes passes a zone characterized by barchans and small star dunes, and there are more small linear dunes and isolated star dunes north of the Sierra del Rosario. The mixed orientations of the dune groups reflect the differences in wind direction induced by the Sierra. Going west, the dunes change slowly into simple small barchans.

Two fault zones constitute

the western and eastern edges of the Gran Desierto and define the basin of the Colorado's delta. A group of faults east of the Gulf of California has established a small scarp that divides the land into a higher elevation 120–180 m (394–590 ft) high and a basin; that governs the abrupt passage from the basin to the Sierra de Juarez. Elevations vary enormously in this territory, even on the local scale. Gradients can be as much as 1,200 m (3,937 ft) in 6.5 km (4 mi), yet the districts west of the basin commonly stand orado River are redistributed since it is responsible for the strong currents (over 2 m/s or 6.5 ft/s) and the deep tidal ranges (more than 10 m/33 ft). The result is that vast tidal flats and barriers develop in a desert environment. Due to aridity, vegetation is thin

the dunes of the Gran Desierto cover the higher segment. The western edge of the basin is a normal fault about 700 m (2,300 ft) high. The Gulf of California's elongated structure determines how deposits from the Colorado and salt deposits pile up in the lowest areas of the supratidal plain.

Mojave Desert

Nation:	United States
Expanse:	65,000 sq km (25,096 sq mi)
Average annual temperature:	38°C (100.4°F)
Rainfall:	<130 mm (<5 in)/year

Facing page: Giant cactus 10 m (33 ft) high and taller characterize the Mojave Desert.

The Mojave Desert is a zone of transition between the hotter Sonora and the cooler, topographically elevated Great Basin deserts. As a transitional feature, it is sometimes considered an integral part of the Sonora Desert and other times of the Great Basin. Its confines are thus rather vague, but it can be said to cover a surface of 65,000 sq km (25,096 sq mi) in the states of Nevada, Arizona, Utah and California. The Mojave's topography is typically expressed in mountainous highlands and basins sheathed in thin vegetation. Wide valleys with gravel- and sand-covered floors drain toward salt flats where borax, potassium and salt are mined. The Mojave stretches from the Sierra Nevada on the west, the Colorado Plateau on the east, and the San Gabriel-San Bernardino Mountains and Colorado Desert on the south. Death Valley—the lowest elevation in the United States—is on the border of the Great Basin and the Mojave. The climate of the Mojave is typified by high daily thermal excursions and by less than 130 mm (5 in) of average annual precipitation. Rainfall is concentrated in winter when temperatures may drop below freezing. Summers are hot, dry and windy.

Though the Mojave can be considered indistinguishable from adjacent deserts, it has a few unique characteristics. The most salient is an abundance of endemic plant species that are exclusive to this desert. Cactus is normally limited to the coarse terrains of the *bajada* (a merger of several alluvial cones to create a continuous fascia of detritus). Two species of yucca, the Mojave yucca (*Yucca schidigera*) and the Spanish Bayonet (*Yucca whiplei*), are more profuse at higher elevations. Common bushes in the Mojave include creosote, shadscale (*Atriplex*), bladder-sage (*Salazaria*) and blackbush (*Coleogyne*). Occasional acacias grow along the arroyos (wadis). Unlike the Sonora, both the number and species of trees are scarce here, though the Joshua tree (*Yucca brevifolia*) is one exception. This strange, tree-like yucca is normally considered the first beacon of this desert's characteristic plant life.

The Mojave contains many playas whose origin is tied to the singularly arid conditions that cause river and soil-content water to evaporate to the point of depositing a salt crust on the surface. Soda Lake starts where the Mojave

River dissipates in the desert. The lake only contains water in humid years and its waters may lie very close to the surface in drier years. Capillary action carries water upward where it evaporates, leaving a white crust of minerals like carbonate and bicarbonate of soda. Surface mirages are a common phenomenon when the lake is dry. When the cli-

mate was more humid and numerous pluvial lakes occupied the whole Basin and Range Province, a huge perennial lake overlaid many of the current playas in this area until 8,700 years ago.

Dunes in the Mojave can reach 160 m (525 ft) high, as in the Kelso Dune complex. These dunes are only a small part of the great sand trans-

port system that includes the flat area known as the Devil's Playground.

Bristol Dry Lake is currently exploited for salt production, and its reserves amount to 60 million tons. Normal cooking salt is found here in the form of crystal rock salt growing under the surface of the playa, which is itself made of clay, sand and chalk. This surface material is thinner than it is in the southern zone, where it is less than 1 m (3 ft) thick, and thicker in the northern zone, while the layer of rock salt is about 1.5 m (5 ft) thick. Deep down, other layers of rock salt are interspersed with layers of clay and volcanic ash. A few centimeters of water cover the lake in winter, but most of the year it is dry and acutely salty. Gypsum and calcium chloride are also produced here intermittently.

Great Salt Lake Desert

Nation:	United States
Expanse:	50,000 sq km (19,300 sq mi)
Average annual temperature:	32°C (90°F)
Rainfall:	<100 mm (<4 in) /year

Right: The surface of this ancient pluvial basin is now a thick crust of salt laid out in polygonal geometric shapes.

Facing page: Earth pinnacles, peculiar erosional structures destined with time to disappear forever.

The Great Salt Lake Desert developed in the same basin as the ancient Lake Bonneville and is part of the Great Basin Desert. Lake Bonneville was one of the biggest pluvial lakes in North America and its remnants are today's Great Salt Lake, Utah Lake and Sevier Lake.

The Wasatch Mountains mark the boundary between the Great Basin and the Middle Rocky Mountain Province. The pediment border of these mountains coincides with a zone where a "normal" fault trails for several kilometers. This fault involves recent glacial deposits and is a structure from the Late Cenozoic; even though it is not still active locally, it experiences slight seismic movements.

At Lake Bonneville it covered 51,700 sq km (19,960 sq mi) and was 335 m (1,100 ft) deep. Geologists have analyzed over 307 m (1,007 ft) of bottom sediments from the ancient lacustrine basin in detail, covering the environmental variations of the last 3 million years. The lake has dried out 28 times in the last 800,000 years, and it is

The desolate landscape of the American desert.

thought that these conditions were associated with a hot interglacial climate. The level of the lake was high during cold intervals—during the Ice Ages—when precipitation exceeded evaporation. The Great Salt Lake's shoreline now rests at an elevation of 1,285 m (4,215 ft), even though local climatic variations cause periodic fluctuations. Multiple shorelines identified for Lake Bonneville indicate that in the

terized by the sudden evacuation of the lake, which dumped enormous quantities of water into the Snake River within a few weeks, causing the shoreline to drop 100 m (330 ft) and remain at this level for about 2,000 years. Later, the level of the lake gradually declined until 11,000 years ago when its shoreline was only 9 m (30 ft) above the current level of the Great Salt Lake.

Lake Bonneville was not the only pluvial lake in the Great Basin. During the last Ice Age, dozens of lakes developed in almost every depression in the Great Basin. The pluvial lakes depended upon the combined influence of increased precipitation and decreased evaporation, lower temperatures, and the local apportionment of glacial-melt waters from valley glaciers in the Rocky Mountains. The southward movement of the western jet stream due to the immersion of the northeastern United States under the Laurentian Ice Cap may have induced these processes.

last 150,000 years, there have been two great lacustrine episodes. The first of these ended about 130,000 years ago, while the second began about 25,000 years ago, during which the surface of Lake Bonneville reached an elevation just over 1,550 m (5,085 ft), the lake's maximum level, about 16,000 years ago. The lake maintained this level for about 1,500 years. The end of this lacustrine phase was charac-

Great Basin Desert

Nation:	United States
Expanse:	490,000 sq km (189,190 sq mi)
Average annual temperature:	27°C (81°F)
Rainfall:	180–300 mm (7–12 in)/year

The Great Basin is the largest desert in the United States and covers an arid territory of 490,000 sq km (189,190 sq mi). The Great Basin contains more than 150 single desert basins, some of which have their own names such as High Desert, Low Desert, Smoke Creek Desert, Black Rock Desert, Great Salt Lake Desert, Sevier Desert, Escalante Desert and Amargosa Desert. More than 160 small mountain groups separate them from one another. All but three of the basins are of the closed variety. The Sierra Nevada on the west, Rocky Mountains on the east, Columbia Plateau on the north and the Mojave and Sonora deserts on the south border the Great Basin. According to some classifications, these last two deserts are an integral part of the Great Basin. Geologically, the Garlock Fault divides the Great Basin from the Mojave. It is a large fault and still active, but the three deserts are largely differentiated by their own characteristic vegetation.

The Great Basin is characterized by a classic basin-and-range topography, with elongated mountain chains that generally run north-south and defined by faults that alternate with arid valleys whose floors are filled with lacustrine and alluvial deposits. Extensive tectonic movements affect the entire area. Because of the region's tectonic activity and aridity, water drainage is almost com-

pletely endorheic. Relatively short streams terminate in saline flats or in playa lakes. Even the Humboldt and Carson Rivers end in a depression known as the Carson Sink. Recent tectonic activity has resulted in large topographic gradients.

It is not uncommon for a mountain chain and an adjacent valley to differ in eleva-

Agaves, a source material for tequila, is typical of the United States landscape in this area.

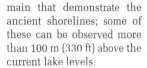
tion by 2,000 m (6,560 ft). Asymmetrical mountain slopes generally characterize, the landscape of the Great Basin. Relatively long rivers run on one flank of the chain and their slope is generally mild, while short, steep watercourses characterize the opposite slope. For example, in Death Valley extensive allu-

vial cones up to 10 km (6 mi) long develop at the base of the western flank, while cones on the eastern flank are not much more than half a mile long. The disparity in slopes is the consequence of asymmetric down-drop of the grabens forming the valley.

The region's aridity is due to the Sierra Nevada, which thwarts the penetration of humid ocean air into these areas via a mechanism, similar to the Alpine föhn, but known as a chinook in this area. This occurs in a "cold" desert due to both high latitude and very high altitudes. Precipitation ranges from 180–300 mm (7–12 in) a year and is more evenly distributed through the course of the year than it is in other North American deserts. Precipitation in the form of snow is common in winter.

The Great Basin has not always been so arid. On numerous occasions during the Pleistocene, its valleys were occupied by vast lakes, which are documented by the extremely flat floors of the basins themselves. Old terraces re-

main that demonstrate the ancient shorelines; some of these can be observed more than 100 m (330 ft) above the current lake levels.

The most economically important playa in the Great Basin is 100-sq-km (39-sq-mi) Searles Lake. Trona and a wide variety of other valuable salts are presently extracted from it. A 930-m (3,050-ft)-long sediment core from the middle of the lake allowed researchers to identify the climate modifications of the last three million years.

Unlike the other deserts, plant life in the Great Basin is scarce and homogeneous, so much so that it may be dominated for kilometers by the same species of bush. Yucca and cactus are rare. Playas, or salt flats where virtually nothing lives, are one of its predominant ecosystems. Only halophile plants exist on the playa, along with insects and marine crawfish in their salt lakes. These ecosystems only diversify at higher altitudes where subalpine forests grow. One of these hosts the oldest tree in the world, "the old Methuselah," which is a *Pinus aristata* more than 4,000 years old.

229

Chihuahua Desert

Nation:	United States/Mexico
Expanse:	518,000 sq km (200,000 sq mi)
Average annual temperature:	38°C (100°F)
Rainfall:	<250 mm (<10 in)/year

The sharp peaks of the eastern Sierra Madres mark the eastern border and edge of the Chihuahua Desert.

The Chihuahua is the eastern- and southernmost of the great North American deserts and is the largest of them at 518,000 sq km (200,000 sq mi). In Mexico, the Chihuahua is located within the Central Plateau, bound by the western and eastern Sierra Madres. The Zacatecas region forms its southern border. The northern part of the desert extends into southwest Arizona, southern New Mexico, and into Texas as far as the Pecos River.

Given its inland position on the continent and relatively high elevations that range from 600–1,675 m (1,970–5,500 ft), the Chihuahua tends to have hot summers and cool-to-cold winters. The whole area is subject to occasional winter freezes, while its northern portion freezes regularly in winter. Most of the area receives less than 250 mm (10 in) of precipitation per year, which is generally concentrated in the summer months. At the end of the summer, in fact, strong storms arrive from the east and may dump 25 mm (1 in) of rain an hour. This rainfall launches wide-

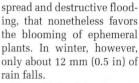

spread and destructive flooding, that nonetheless favors the blooming of ephemeral plants. In winter, however, only about 12 mm (0.5 in) of rain falls.

The middle of the desert is characterized by *bolsons*, or

endorheic basins, remnants of the ancient pluvial lakes, which sometimes host playa lakes, locally known as "lagunas." There are thermal springs in the Bolson Cuatro Cienegas (or Four Lakes Basin) that are rich in sulfates and other dissolved minerals from the Sierra de San Marcos, which divides the bolson in two. When lake water evaporates during arid spans, as in the Laguna Grande, gesso crystallizes at the edges of the lagunas.

The prevailing eastern wind picks up the crystals and accumulates them into dazzling white mobile dunes as high as 6–9 m (20–30 ft). This chalk can also form far-reaching, dense grayish crusts.

The Rio Casas Grandes runs across the northern area of the desert in Mexico and into the desert where it branches into braided channels before

Above: The typical appearance of rocks eroded by the extreme flooding of rare but intense rains.

Facing page: The eastern Sierra Madres, edge of the desert.

emptying into the Guzmàn Laguna. The Casas Grandes starts in the western Sierra Madre where more than 430 mm (17 in) of rain typically falls in summer alone. This is one of the rare playa lakes with standing water all year long. Even though the river drains a 16,00-sq-km (6,178-sq-mi) basin and the lake receives more spring- and rainwater, evaporation is so great that Guzmàn Laguna is never deeper than 1 m (3 ft), and generally much less so, yet it is typically 8 km (5 mi) long and 4 km (2.5 mi) wide. Like all bolson lakes, the laguna has no tributaries so the sediments that the river deposits in the lake accumu-

late in place and bury the laguna. Eventually, only a dry, sandy plain will remain.

In dry bolsons, summer storms transform the surface into mobile sands. When evaporation dries out the muck, desiccation fissures that measure up to 3 m (10 ft) develop, forming a honeycomb-like pattern of cracks. The phenomenon involves vast surfaces. The winds, which blow from the west in the northern part of the desert, entrain the dust particles and redistribute them on the desert. Only in a few sheltered areas east of the bolsons, which act as dust traps, do medanos or dunes form to heights of 100 m (330 ft). This enormous area is very diverse geologically, but in general, it can be said that limestone is the most widespread rock.

Cold-sensitive plants are localized at low elevations between the Bolson de Mapimi and the Big Bend Region in Texas. At higher elevations and latitudes, however, these species decline in number and then disappear completely.

Unlike the Sonora Desert, which is widely characterized by large cacti and small trees, the Chihuahua is essentially a shrub desert characterized by the rather modest biological diversity of its vegetation. The creosote (*Larrea tridentata*), with its penetrating odor of resin, is found almost everywhere, while the tarbush (*Flourensia cernua*), one of this desert's typical species, is sporadic but can form extended brush "forests" if soil conditions and humidity permit.

Common plants in the north include the four-wing saltbush (*Atriplex canescens*), mariola aster (*Parthenium incanum*) and mesquite (*Prosopis glandulosa*). Succulents are represented by medium-sized cacti, yuccas (*Yucca elata, Yucca Torreyi*) and agave. Among these latter the shindagger or lechuguilla (*Agave lecheguilla*) is wide-spread and only grows in the Chihuahua Desert. It has a thin stem 4–5 m (13–16 ft) tall, which bursts with multicolor flowers when it blooms and emerges from a rosette of basal leaves characterized by terminal spines that are sharp enough to puncture boots.

As in the Sonora Desert, animal life is abundant in the Chihuahua Desert: rabbits (*Sylvilagus audubonii*), hares (*Lepus californicus*), cactus mice (*Peromyscus eremicus*), foxes (*Vulpes velox*), rattlesnakes (*Crotalus scutulatus*) and the greater roadrunner, known to all as the famous "beep-beep" of animated cartoons (*Geococcyx californianus*). The roadrunner, which is also found in the Sonora Desert, can run as fast as 40 km/h (25 m/h). Strangely, it uses fewer muscles to run than to fly and thus realizes an energy savings that is indispensable for life in the desert.

Atacama Desert

Nation:	Chile
Expanse:	200,000 sq km (77,220 sq mi)
Average annual temperature:	14°C–30°C (57°F–86°F)
Rainfall:	not measurable (2 mm (0.08 in) in the last 30 years)

a mountain chain character-ized by altitudes 1,000–1,500 m (3,280–4,920 ft) above sea level, peaks to 2,000 m (6,560 ft), and mountainsides that drop precipitously to the ocean. The eastern border is the Cordillera Domeyko, which precedes the impres-sive Andes Cordillera. The desert has an average eleva-tion of 900 m (2,950 ft) and is composed of sands, erosional gravel and volcanic matter from the Andes chain.

This material accumulates between the Coastal and An-

Above: Typical vegetation of the Chilean desert with active volcano Lascar in the background.

Right: The salt-rich surface produced by the intense evaporation in this locale.

The Atacama is a coastal des-ert in northern Chile approx-imately between the provinces of Antofagasta and Atacama, south of the River Loa and north of the Copiapò river basin. On the north, it ex-tends through the Tarapaca region all the way to the Peruvian border. The desert appears narrow and elongated north to south with a maxi-mum length of 1,000–1,100 km (622–684 mi). Along the Pacific Coast to its west lies the Cordillera de la Costa,

des cordilleras, filling the tectonic depression between the two chains. Its surface is characterized by deposition cones laden with rounded stones from the Andes and wide hollows covered in saline crusts known as playas, which are all that remain of ancient lacustrine basins. Because the diurnal and seasonal winds that affect this region are so variable, the wind-transported material does not form sand dunes but disperses uniformly. Dunes can only be found occasionally along the coastal side of the desert and somewhat more rarely inland from the Cordillera de la Costa. The morphology is predominantly created by powerful forces of wind erosion and evaporation, which result in the myriad bizarre shapes assumed by wind-scoured rocks and playas.

The rivers that flow from the Cordillera de la Costa toward the ocean cut deep rock canyons called *quebradas*, which may reach depths of 800 m (2,625 ft) near the coast. The rivers that descend to the east, instead, cut into the central depression until they arrive at the foot of the Andes.

The Atacama contains extremely rich saline mineral deposits such as the sodium nitrate that has been mined in this region since the middle 1800s for the manufacture of explosives and fertilizers. These deposits are found just about everywhere in the desert and were transported by torrents and underground waters that flowed from the numerous Andean volcanoes and

and the protection from humid ocean currents furnished by the Cordillera de la Costa render the Atacama Desert the most arid place on earth. Along the coast, however, its aridity is also a consequence of the cold Peruvian current.

By causing a thermal inversion (cold stable air on the ocean surface and warmer air higher up), the current pro-

Above: Rocky spires in precarious balance.

Right: The Andes Cordillera borders the northern region of the Atacama Desert.

came to rest in the central depression, forming great salt lakes. The arid climate, which has endured for almost 15 million years, provoked the constant evaporation of water and consequent accumulation of salts that permitted the formation of thick evaporite deposits. The high-pressure cells of the southern Pacific

vokes dense fogs and stratus or sheet clouds, but no rain. As a consequence, many places on the desert have not recorded measurable rains since measuring began, and in the coastal city of Iquique, only 2 mm (0.08 in) of rain has fallen in 30 years. Temperatures are relatively low, and the daily thermal range is 0°C–30°C (32°–86°F). North-west of Antofagasta in the heart of the desert is the city of Calama on the banks of the Rio Loa. It has seen huge growth since the middle 1800s due to the mining of abundant copper and sodium nitrate deposits in the region. The Tatio geysers are farther south and east, on the desert's east-central border near San Pedro de Atacama at the foot of the Andes. They occupy an emblematic area where dozens of fumaroles emit vapor into a landscape of desolate stone figures carved by the wind over the millennia.

AUSTRALIA

Australia is the most arid continent, excluding the Antarctic, with a total of about 5.4 million sq km (2.1 million sq mi) subject to desert conditions. Its aridity is nowhere extreme, however, and average annual rainfall never drops below 100–125 mm (4–5 in).

Given the prolonged tectonic stability of vast areas in Australia, different geomorphologic features have been inherited from each of the great variety of climates the continent has experienced since the Jurassic, and probably even earlier. Much of the flat and weakly undulating desert landscape dates to the Cretaceous Period. Many of its plateaus, mesas and great lacustrine depressions are from the Tertiary, while its sand dunes and numerous small playas are Pleistocene. The Holocene had little impact on contemporary deserts, even if the Europeans' important contribution to the degeneration of the ecosystems in certain arid and semiarid zones must be taken into consideration.

About 50–60 million years ago Australia began to separate from the Antarctic, migrating north and approaching its cur-

INDONESIA

47. Great Sandy Desert

48. Simpson Desert

Tropic
of Capricorn

AUSTRALIA

49. Great Victoria and Gibson Deserts

Indian Ocean

TASMANIA

rent latitude during the Miocene (about 20 million years ago). Australia has thus suffered climatic variations due to its change of latitude, but has also experienced the effects that such movements had on the circulatory, climatic and oceanic systems in the southern hemisphere, and the climatic global changes associated with generally lower temperatures during the Cenozoic. Among Tertiary relics in the Australian desert, there are widespread duricrusts containing silcretes and laterites, and paleodrainage systems stretching across vast erosional surfaces. In addition, there was a landlocked sea in the Lake Eyre depression.

A more arid climate began to install itself in the Upper-Miocene-Pliocene.

The huge Australian sandy deserts were created in the Pleistocene. There is evidence that the linear dune fields for which Australia is famous stretched far beyond the current boundaries of the desert areas. The dunes on the northern part of the Great Sandy Desert extend below the Holocene alluvial deposits of the Fitzroy Estuary and were thus formed in periods of low sea level during the Pleistocene.

All the data from sedimentary core samples taken from both the continent and the ocean show that during the last glacial maximum (about 25,000–18,000 years go), much of Australia was drier and windier than it is today and was surrounded by wider continental shelves. Seventeen thousand years ago, the temperatures, rainfall, and sea level began to rise, and most of the desert dunes stabilized about 13,000 years ago. Rainfall increased at the beginning of the Holocene, bringing about the extant growth of vegetation and forests.

Great Sandy Desert

Nation:	Australia
Expanse:	360,000 sq km (139,000 sq mi)
Average annual temperature:	23°C–37°C (73°F–98.6°F)
Rainfall:	>250 mm (>10 in)/year

Above and right: From dunes of red sand to rounded granite sculptures hundreds of millions of years old, the scenic variety of Great Sandy Desert is enormous.

The Great Sandy Desert in northwest Australia is a vast expanse of sandy hills and briny bogs, lying within the huge Canning sedimentary basin. The Great Sandy extends from Eighty Mile Beach on the Indian Ocean to the west, to the Northern Territories border on the east, adjoining the Gibson Desert on the south and the mountainous Kimberley area on the north. Like many desert areas around the world, almost half of this territory is composed of sand, with one part being dominated by aligned dune fields while the other has wide plains covered by a thin layer of sand, over which a delicate mantle of vegetation grows.

Linear dunes do not migrate in particular directions, as often happens in other areas, and their stability is probably due to the vegetation that often covers and colonizes them. For this reason, many prefer to define them as sandy crests rather than true dunes. The crests are generally 40–50 km (25–30 mi) long and about 15 m (50 ft) high. The west-northwest trend of the Great Sandy's linear dunes is also seen in the Simpson Desert to the southeast. The winds blow constantly in the same direction in both deserts. The directions of the linear dunes and therefore of the dominant winds in Australia

Huge accumulations of sand tinted red by its high concentration of ferrous oxides are this desert's most common landscape.

maintain a very specific bearing that follows a large clockwise spiral within the continent. Even if there are no clear data that indicate with precision the era of these dunes' formation, a few

surveys and datings established that they are from the Lower Pleistocene, or about 125,000 years ago.

Some scholars hypothesize that the present structure of the desert, which has

Australian deserts, rainfall is scarce, even though it seems rather high in relation to arid zones in the rest of the world. In its driest areas, in fact, average annual rainfall rarely falls below 250 mm (10 in). Monsoon-related storms and tropical cyclones cause this "abundance" of water. Along the coast, in Kimberley's extreme north, the annual averages exceed 300 mm (12 in) a year, but it is compensated by an extremely high rate of evaporation. Diurnal temperatures measured in the summer season are among the hottest in Australia and hover between 37–42°C (97–108°F); temperatures in the short winter are also fairly high at 25-30°C (77°–86°F).

not varied much over time, was caused by winds connected to an arid period due to a glaciation that happened elsewhere. This is confirmed by the fact that groups of dunes along Eighty Mile Beach extend at least 15 m (50 ft) into the ocean and were submerged when melting glacial waters raised the sea level. Many plants that take advantage of the humidity stand out. As in other

Simpson Desert

Nation:	Australia
Expanse:	300,000 sq km (115,830 sq mi)
Average annual temperature:	summer 37°C (100°F); winter 20°C (68°F)
Rainfall:	150–200 mm (6–8 in)/year

Above and right: The ancient sandstone structures of the Olga Mountains rise a few kilometers from Ayers Rock in the boundless desolation of the Northern Territories.

Situated in the heart of the Australian continent, the Simpson Desert lies in the southeastern corner of the Northern Territories, bordering with Queensland and South Australia, while on the west it is bordered by the Finke River. The Finke carves a corridor among the desert's red rocks and is the oldest river in the world still flowing in its original direction. Most of its course lies in the Finke Gorge National Park about 138 km (85 mi) south of Alice Spring. The northern border of the Simpson is marked by the Mac-Donnell Mountains and the River Plenty, by the great salt Lake Eyre in the south, and finally by the Mulligan and Diamantina Rivers on the east.

The desert is characterized by long parallel dunes of red sand, similar to those found in the nearby deserts of Great Victoria and Great Sandy, whose characteristic color is due to the ferrous oxide that covers every single grain of sand. Regional winds blow northwest-southwest, causing the dunes to reach an uninterrupted length of 290 km (180 mi), a height of 38 m (125 ft), and a width of 300 m (980 ft) in certain large tracts. It was recently discovered that these migrate slowly north and northwest at an

average of 9 m (30 ft) per year. At that rate, the great Eyre Basin's wind-transported lacustrine and alluvial deposits may have constituted the present Simpson Desert during the Lower Pleistocene (about 130,000 years ago). The lowest dunes in the western part of the desert prove to be farther apart in relation to the highest dunes in the eastern zone. The troughs between one dune and another are fairly regular in width and average little over a kilometer. They may be occupied by

playas, sand flats or stone masses in the eastern zones that are laden with ferrous minerals and dominated by powerful eolian erosion. Despite the fact that they received less than 250 mm (10 in) of rain a year and lose even more in evaporation, rare inundation by waters from the innumerable basins that surround the lake itself. There are also saline troughs in the same zone, which often contain chalky incrustations that form when the water in the small lakes evaporates completely. Lake

Below and below right: Ayers Rock and the Olga Mountains are surely among the best known symbols of the Australian continent.

the playas cannot be considered lifeless, given their intense colonization by plants and animals. The small clay flats are especially common in areas adjacent to the basin of Lake Eyre and are created in the wake of Eyre, which provides the desert's southern border, covers 9,300 sq km (3,590 sq mi) and sits 11 m (36 ft) below sea level. Much of its surface is a salt flat that fills with water only twice every 100 years when the rivers flowing from Queensland receive enough water to reach the heart of the desert. It is one of the biggest salt flats in the world. This immense territory harbors innumerable natural wonders, including the famous Henbury Meteorite Craters

Australia with an average annual precipitation of 150–200 mm (6–10 in) concentrated in summer and autumn. Due to seasonal monsoons, those rains assume a torrential character, dropping the rain of three to four years all at once. Daily

104°F) and peaking at 50°C (122°F). Sandstorms are common in the hottest periods.

Conservation Reserve, a small park 147 km (91 mi) southwest of Alice Spring, which contains 12 craters created millions of years ago by the impact of a meteorite.

The Simpson Desert is located in the driest areas in

and seasonal thermal fluctuations are significant. In winter, diurnal highs hover between 18–23°C (64–73°F) and sometimes drop to 0°C (32°F). Summer daytime temperatures are very high, averaging 35–40°C (95–

Great Victoria and Gibson Deserts

Nation:	Australia
Expanse:	Great Victoria: 423,751 sq km (163,610 sq mi)
	Gibson: 155,530 sq km (60,050 sq mi)
Average annual temperature:	20°C–35°C (68°F–95°F)
Rainfall:	200–250 mm (8–10 in)/year

Above: Bright dunes of white sand gave the place its name, White Sands.

Facing page: Limestone rock pinnacles unequivocally document an ancient climate change.

The Great Victoria and its mobile dunes stretch 1,300 km (808 mi) across southwestern Australia, covering an area of about 420,000 sq km (162,162 sq mi). Since it mixes with the Gibson Desert on the north and with the Nullarbor Plain on the south, the exact surface of the Great Victoria is poorly defined. Explorer Ernest Giles dedicated the desert to Queen Victoria when he crossed it for the first time in 1875.

Great Victoria contains many extensive Aboriginal reservations; agriculture is scarce and limited to the edges of these deserts. Rainfall is low at 200–250 mm (8–10 in) per year, but is relatively abundant in relation to other deserts. Rainfall is extremely variable, especially in the Gibson Desert. At its southern edges near Leonora, rainfall increases thanks to meteorological disturbances that augment winter rain loads. The Gibson's domain also depends on the activity of tropical low-pressure regions and, occasionally, upon extra-tropical cloud masses. Lows are frequent, numbering 20 to 30 a

year in the Gibson and averaging 15 to 20 a year in the Great Victoria.

Daytime summer temperatures at elevations above 500 m (1,640 ft) range between 32–40°C (90–104°F). Winter temperatures are mild to hot, averaging 18–23°C (64°–73°F), but the winter is short and by late September temperatures have already returned to high levels. Meteorological disturbances in the southernmost area cause lower temperatures to be recorded for several days each winter. Freezing is not rare, but even common in some parts of the Great Victoria.

The desert is about 200 m (650 ft) above sea level in the south and about 700 m (2,300 ft) in the north. This slope represents an ancient erosional surface cut from Archeozoic rock and covered in eolian sands. Longitudinal dunes with an easterly orientation nearly cover the entire surface of the desert. The dunes are about 20 m (65 ft) high and often longer than 100 m (330 ft). Numerous granite inselbergs are surrounded by red alterites, and there are many small salt lakes in the central zone. Sandy flats and the interdune areas to the west are occupied by a typically bushy and arboreal vegetation with abundant eucalyptus. The inselbergs also

248

harbor vegetation. The biggest inselbergs can capture rainwater and channel them along paths of preferential flow that run a certain distance in the desert before creating small stretches of clay or small saline lakes. The vegetation is almost totally local in origin and this is a good indicator of the sive and abundant to support regular fires. Lightning strikes during the summer cyclones usually trigger the fires, and they may last several weeks. It is impossible to forecast the reach of these fires since they involve indifferently broad areas covered by vegetation or small plant communities. The factors sandstone in the Canning Basin. The material is left over from an ancient soil that developed in a hot, humid climate very different from the present one. Various bushy and steppe-like plant communities grow on it even now, and there are also small fields of dunes.

value of these deserts as habitats. On the other hand, a few mammals such as rabbits, cats and camels have been introduced in the area and have adapted well. The original fauna includes only a few birds, reptiles and invertebrates.

Spontaneous fires are a dominant factor in the Great Victoria's ecosystem. Its vegetation is sufficiently perva- that influence the fires' frequency, scale and geometry are temperature, rainfall, wind conditions, plant biomass, the plants' spatial distribution and the presence of natural firebreaks. The fires have been studied and monitored with aerial and satellite photos for many years.

The Gibson Desert is a laterite plain that covers the Cretaceous and Jurassic

M. Cremaschi, S. Di Lernia
Wadi Teshuinat Paleoenvironment and
Prehistory in South-Western Fezzan
University Center for Research on the
Civilizations and Environment of the
Ancient Sahara, 1998

R.A. Bagnold
Libyan Sands
London: Hodder

R.A. Bagnold
The Physics of Blown Sand and
Desert Dunes
London: Methuen

R. Biasiutti
Il paesaggio terrestre (The Earthly
Landscape)

M.E. Brookfield—T.S. Ahlbrandt
Aeolian Sediments and Processes
Elsevier, Amsterdam, 1983

G.B. Castiglioni
Geomorfologia (Geomorphology)
UTET

K. Cloudsley—J. Thompson
I deserti (Deserts)
De Agostini Geographic Institute,
Novara

H. Cuny
Les desert dans le monde
(Deserts Around the World)
Paris: Payot, 1985

M. Derruau
Precis de geomorphologie
(Digest of Geomorphology)
Paris: Masson

L.E. Frostik—I. Reid
Desert Sediments: Ancient
and Modern
Blackwell Scientific Oxford, 1987

K.W. Glennie
Desert Sedimentary Environment
New York: Elsevier

D.W. Goodall—R.A. Perry
Arid Land Ecosystems
Cambridge University Press, 1979

Brian John
L'evoluzione del paesaggio
(Evolution of the Landscape)
Novara: De Agostini Geographic
Institute

M. R. Leeder
Sedimentology
Allen and Unwin

H.H. Lettau—K. Lettau
Exploring the World's Driest Climate
Institute of Environmental Studies
University of Wisconsin, 1978

J.A. Mabbut
Desert Landforms
Cambridge: MIT Press, 1997

W.M. Mc Ginnies
Deserts of the World
Tucson

W.G. Mc Ginnies—B.J. Goldman—
P. Paylove
Deserts of the World
Tucson: University of Arizona Press,
1980

Martinus Nijhoff
Desert and Arid Lands
Den Haag: El Baz, 1984

Van Nosmand Reinhold
Encyclopedia of Geomorphology
R.W. Fairbridge, 1968

Mario Panizza
Elementi di geomorfologia
(Elements of Geomorphology)
Bologna: Pitagora Editrice

F. Press—R. Siever
Earth (Italian version Introduzione
alle scienze della terra, Zanichelli)
W.H. Freeman & Co.

Raymond Siever
Sabbia (Sand)
Zanichelli editore, 1990

A. Starker Leopold/Time-Life
Il deserto (I regni della vita) / The
Desert (The Kingdoms of Life)
Mondadori, 1962

A.N. Strahler
Geografia fisica (Physical Geography)
Piccin

W.D. Thornbury
Principles of Geomorphology
John Wiley & Sons

D.J. Yaalon
Paleopedology: Origin, Nature and
Dating of Paleosols
Jerusalem: Israel University Press,
1971

S. and J.G. Zotl
Quaternary Period in Saudi Arabia
Vienna: Al-Sayari, 1978

D.L. Weide
Soil and Quaternary Geology of the
Southwestern United States
Geological Society of America, Special
Paper 1985

K. Wolton
The Arid Zone

Albedo
Reflective power of a surface; the earth's overall average represents 35% of the solar radiation that hits the planet.

Anticyclone
Areas of atmospheric pressure higher than surrounding areas. In an anticyclone, wind moves from the center toward the perimeter with a clockwise rotary motion.

Biome
Aggregation of plants and animals formed by environmental factors, such as climate, the nature of the terrain and so forth.

Cyclone
A system of low atmospheric pressures in which the wind moves toward the center of the depression with a counterclockwise rotary motion in the northern hemisphere and in a clockwise direction in the southern.
    A tropical cyclone or hurricane is a highly intensified low-pressure system whose winds reach high velocities.
    Tropical cyclones always originate in the same ocean areas and always move in the same direction though on various trajectories.

Craton
Part of the continental interior, a flat and seismically stable area that has not suffered tectonic deformation for several hundred million years.

Depression
A geographic zone resting below the topographic norm.

Diapir
A cupola or mushroom-shaped structure created by the intrusion of plastic and lightweight rock mass upward through denser rocks overhead.

Duricrust
An especially hard and mineral-rich surface element.

Emungement
Extraction of water from the subsoil.

Endorheic basin
Hydrographic basin whose waters do not flow to the sea but are lost to evaporation or terminate in lakes that occupy huge depressions in the continental interior (e.g., Caspian Sea, Aral Sea, Lake Chad).

Fault
Fracture within a rock mass along which a relative movement (dislocation) of the parts may occur. The particular movement may be uplift, a subsidence or a slip.

Freatic waters
Subterranean waters harbored in permeable rocks whose depth is only limited by impermeable rocks, while they are free toward the upper surface, which is called the piezometric surface.

Gondwana (Continent of)
Name for an ancient landmass (continent) that existed about 200 million years ago. The other lands were called Laurasia. Gondwana was an assemblage of Africa, Madagascar, Australia, South America and the Antarctic.

Horizon
Rock surface characterized by well-defined lithologic elements that define its nature.

Inselberg
Rock mass isolated by erosion, which plainly emerges from a surface that is in the process of being leveled. It is generally smooth and rounded.

Internal delta
A fluvial structure characteristic of depressed and arid continental zones in which a single-channel river branches out into many channels, losing water by infiltration and evaporation. The infiltrated waters may coalesce again and resurface in a single channel at the end of the depression. Significant examples are the internal deltas of the Niger and Okavango rivers.

Intertidal plain
Mildly sloping zone between the boundaries of high and low tides.

Mud crack
Structure formed when a polygonal section of mud dries out enough to curl at the edges.

Karstism
An array of forms created by the solvent action of water on rocks. It involves rocks that are directly soluble in water, such as gypsum and rock salt, and rocks soluble in slightly acid waters, such as limestone and dolomites. It includes both surface and subterranean forms (caves).

Orogenesis
The group of geologic processes that lead to the formation of mountain chains.

Pedogenesis
The combination of chemical, physical and biological processes that create soil.

pH
Symbol used to express the concentration of hydrogen ions in a solution. The scale goes from 0 to 14. A pH of 7 represents the neutral value of pure water.

Piezometric level
The pressure in an artesian aquifer causes its water to reach this level when it arrives at a well.

Piezometric surface
The surface that water would reach if an impermeable rock lid were not obstructing its rise and thus its free flow on the surface.

Plateau
High plain generally coincident with structural planes such as the surface of lava strata or flows.

Quartz (SiO2)
Among the most abundant silicates on the planet. Also called silica, which may exist in a non crystalline form.

Sandstone
Sedimentary rock produced by the natural cementation of sand.

Shield
This is the ancient igneous or metamorphic rock nucleus found in cratons flattened by erosion.

Soil
The soil that covers most of the earth's surface is the product of the disintegration of rocks by chemical, physical and biological weathering. The thickness and the degree of soil development are highly variable, and their classification is especially articulated and complex.

Solifluction
Downhill movement of surface deposits saturated with water, especially melt water flowing over a terrain that is still frozen.

Travertine
Sedimentary rock caused by chemical cementation and created by the precipitation of limestone near waterfalls or springs.

Wadi
Also spelled "uadi," a typical watercourse in desert regions. The wadi is almost always dry, leaving its bed visible.

WEB SITES

Deserts: Geology and Resources
(US Geological Survey)
http://pubs.usgs.gov/gip/deserts/

The Ultimate Desert Resource
http://www.desertusa.com/

The National Park Service
http://www.nps.gov/

The Desert Centre: Osoyoos,
British Columbia
http://www.desert.org/

Desert Research Institute
http://www.dri.edu/

The Arid Lands Newsletter
http://ag.arizona.edu/
OALS/ALN/ALNHome.html

Desert Ecology
http://helios.bto.ed.ac.uk/bto/
desertecology/index.htm#home

Biological Soil Crusts
http://www.soilcrust.org/

Deserts of the World
http://www.stemnet.nf.ca/CITE/
desert_world.htm

Virtual Reality Panoramas: Southern
Deserts of California
http://www.virtualguidebooks.com/
SouthCalif/SouthernDeserts.html

Deserts in the World (resources,
books and weblinks)
http://www.naturalenviron.com/
deserts.html

Oxfam's Cool Planet
http://www.oxfam.org.uk/
coolplanet/ontheline/explore/
nature/deserts/deserts.htm

U.S. Environmental Protection
Agency
http://yosemite.epa.gov/oar/
globalwarming.nsf/content/
ImpactsDeserts.html

Cactus Museum
http://www.cactusmuseum.com/

Lonely Planet
http://www.lonelyplanet.com/
theme/deserts/deserts_index.htm

ACKNOWLEDGMENTS

To produce this volume we availed ourselves of several generous collaborators and friends who dedicated their time to us, sacrificing their commitments to research and teaching. So it is with great pleasure and gratitude that we thank all those who were so helpful to us, and especially Dr. Luisa Zuccoli, professor of geomorphology in the Faculty of Geology at the State University of Milan, for her fundamental support in the compilation of certain introductory chapters and in researching data for the desert guides; Professor Mauro Cremaschi, who teaches the geology of the Quaternary period in the Faculty of Geology at the State University of Milan and who is a passionate and engaged researcher, for his compilation of the "Desert Chronicle" chapter; Dr. Luca Tombino, researcher and deputy for the Geopedology Chair in the State University of Milan's Faculty of Geology, who supervised and managed research for the descriptions of deserts provided in the desert guides, which work was undertaken by Francesca Ferraro, Silvia Maggioni, Paola Corti, Marco Salvioni, and Caterina Melandri at the State University of Milan. Without their help, the publication of this demanding book would have been even more exhausting.

One last thanks to Agostino Rizzi of the University of Milan's Earth Sciences Department for making the electron microscope images of grains of sand; and to geologist Andrea Strini; reporter Bruno Zanzottera; Marco Santini; Davide Scagliola; and the Aura photo agency for providing spectacular desert images.